T0076451

TAKING ON BIG PHARMA

DR. CHARLES BENNETT'S BATTLE

Julius Getman
Terri LeClercq

Skyhorse Publishing books may be purchased in bulk at special discounts for sales promotion, corporate gifts, fund-raising, or educational purposes. Special editions can also be created to specifications. For details, contact the Special Sales Department, Skyhorse Publishing, 307 West 36th Street, 11th Floor, New York, NY 10018 or info@skyhorsepublishing.com.

Visit our website at www.skyhorsepublishing.com.

10 9 8 7 6 5 4 3 2 1

Library of Congress Cataloging-in-Publication Data is available on file.

Hardcover ISBN: 978-1-5107-7541-1
Ebook ISBN: 978-1-5107-7542-8

Cover design by Kai Texel

Printed in the United States of America

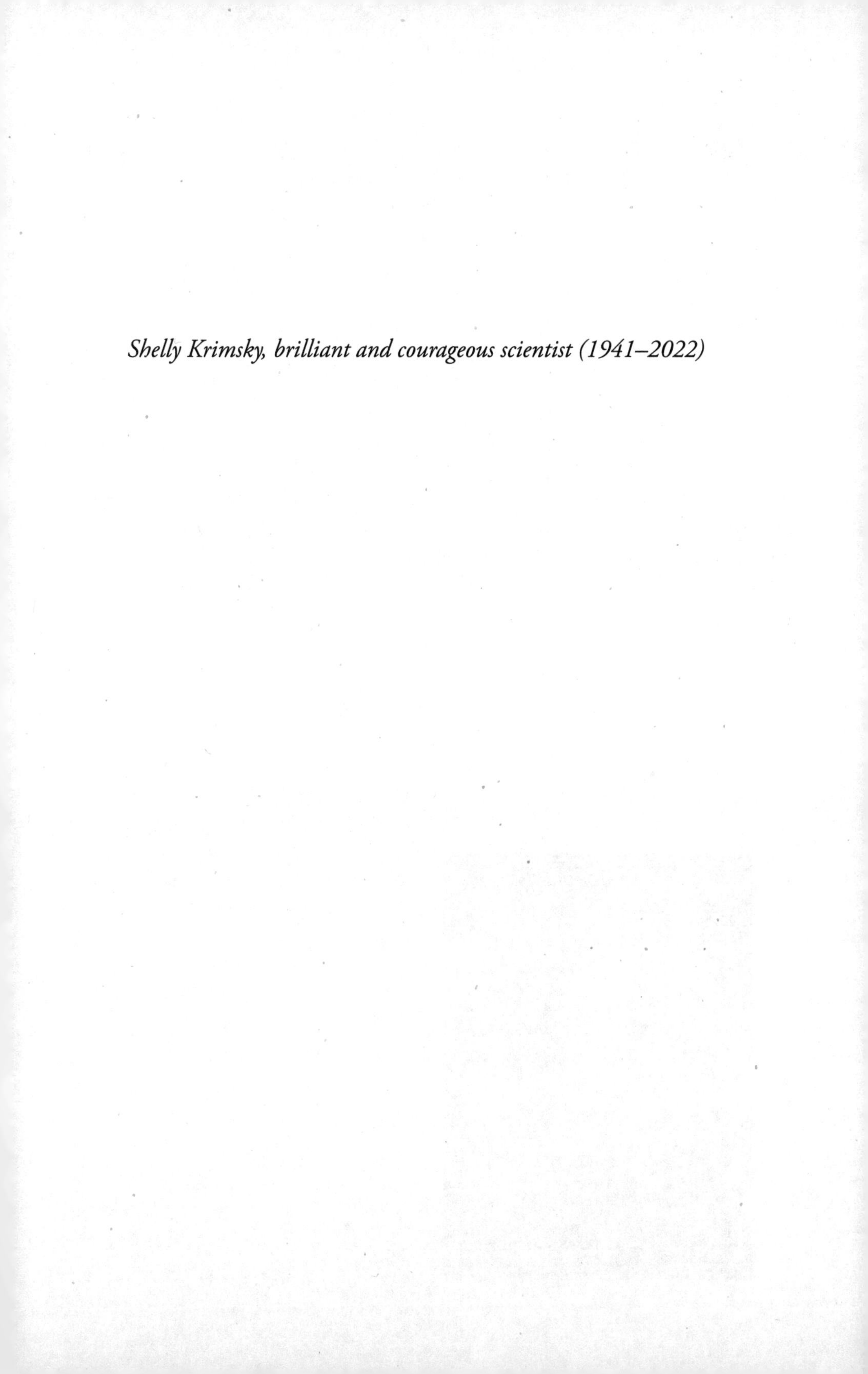

Shelly Krimsky, brilliant and courageous scientist (1941–2022)

Contents

Acknowledgments

If we were to list all the people who helped us,
this thank-you list would be as long as the text.
Still, special thanks to:
Dean Ward Farnsworth, School of Law, University of Texas School of Law;
President Harris Pastides and wife Patricia Moore Pastides, University of South Carolina;
Attorneys including Richard Gonzales, James McGurk, Rob Henning, Peter Winn, Justin Brooks;
Amy Bennett, Charlie's wife and fact checker extraordinaire:
Bloomberglaw.com;
Multiple friends and colleagues of Charlie Bennett who were willing to discuss this controversial issue;
Medical researchers who have investigated and continue to bravely investigate Big Pharma and its tentacles;
Early readers/editors including Kendal Gladish, Brian Saliba, Shermakaye Bass;
Friends and family who endured five years of our talking about erythropoietin and frightening drug reactions.

Acknowledgments

Preface

September 2014. We were on a pleasant cruise, enjoying a Rosh Hashanah dinner.

Moving around the table, each of us introduced ourselves to those sitting nearby, as strangers do. The two people to my right told me their names and home city; one added that he was "in advertising." My wife, Terri, admitted that she was a retired rhetoric professor.

The big guy next to me, speaking in a voice that could be heard at the other end of the table, announced "I'm Charlie Bennett, a well-known academic hematologist oncologist. My career was almost ruined by false accusations. I suspect pharmacy giant Amgen."

Intriguing . . . but perhaps more detail than was appropriate. At my turn, I said I was a retired law professor, and Charlie Bennett interrupted the flow of introductions to announce that a few years earlier he had needed a good lawyer who knew the academic world. I was not interested in Charlie's past legal needs. But during the meal he made a point of interviewing me as though I were a job applicant. "What kind of law did you teach?"

"Mainly labor but also law and literature." I could see his interest fade.

"Where did you teach?"

"University of Texas."

He stifled a yawn. "Where did you go to law school?"

"Harvard."

His face came alive with renewed interest. I had the uncomfortable feeling he was about to ask about my grades. Instead, he rose and shook my hand, smiling broadly. "Hey, why don't we meet tomorrow at ten by the pool bar. I'll tell you my story. I think you'll find it interesting." His wife Amy remained seated with us, obviously not interested in hearing his story again.

I was looking forward to a relaxing and uneventful vacation, but it was hard to refuse; perhaps the story would turn out to be interesting. Next morning, sure enough, Dr. Charlie Bennett showed up with a well-worn folder of articles chronicling both his research and academic successes and his subsequent legal problems. A full folder. At a pool bar. On a cruise.

His large shadow covered my eggs and bagel. His curly hair stood out like a younger Einstein's. "I Googled you last night! You didn't tell me that you taught at Yale and Stanford or that you were general counsel of the American Association of University Professors. I'll bet that you have represented lots of academics in cases that involved allegations by their university employers. This morning I ordered your book, *In the Company of Scholars*—bet I'll get it on the Kindle tonight." His wife Amy had wandered upstairs in time to say, "Yeah, that's Charlie. He always checks credentials." She went over by the windows and watched the waves.

Charlie ordered us screwdrivers and handed me his thick folder: "Look it over and we can discuss it. Ask me any question that you would like." And then as strangers typically do, we shared stories about other travel.

Later that morning back in our room, I studied the articles. Interesting story. Because of accusations that he had violated the False Claims Act, Bennett had been forced to give up a prestigious chair at the Northwestern University's Feinberg School of Medicine and its Kellogg School of Management and accept a joint position at the South Carolina College of Pharmacy and the Medical University of South Carolina in Charleston. His national reputation as a scholar and cancer researcher had been severely tarnished by serious public accusations of misused grant money. I couldn't tell whether Charlie had been totally honorable, foolish, or corrupt, but it was clear that his past was marred by an academic tragedy. I had dealt with many cases of faculty accused of academic misconduct; I thought it highly likely that something he had done or failed to do had given cause for suspicion.

When we met again that afternoon, I asked as subtly as I could whether he had been "mistaken" in submitting grant expenditures—the subject of many of the articles that condemned him. He insisted vehemently that he had never spent a dollar improperly. Perhaps that was why he carried this folder around—to help demonstrate his innocence? If he was telling the truth, the accusations were false and had produced a major miscarriage of justice. I was curious, intrigued . . . might be interesting to follow up. I figured that since I had just retired and was no longer teaching, learning the details of this mystery and writing the story up for him would not take too long. I told my wife Terri about our conversation. She looked up from her mystery novel and responded: "Worth exploring, but it'll be all uphill."

She turned out to be prescient. Uphill was an understatement. Usually, lawyers and university officials are willing to tell their story. But not in Charlie Bennett's case. Confusion, contradiction, and refusal to cooperate by key witnesses and Northwestern University officials thwarted my many in-person, email, and telephone attempts to learn the underlying facts. Of course, Charlie was eager to fill in missing facts. But I had long been aware that even innocent people were likely to present an unreliable version of the facts. I needed to locate and talk with reliable witnesses, including the lawyers Charlie had retained in some of this legal mess.

I was encouraged when Charlie told me that he had been represented by Chicago attorney David Stetler. Surely Stetler would know both the strength and weakness of both Charlie's case and the case against him. Early in 2015 I called Stetler. He insisted that he could not speak to me without Charlie's permission, adding that "If Charlie grants permission for me to talk with you, he'll be making the biggest mistake of his life." Well, that was sure an odd and unexpected beginning. I immediately reported Stetler's reaction to Charlie, who called Stetler the next morning and waived any attorney-client privilege. The next day, Stetler once again refused to talk, alleging an unexplained "conflict of interest."

Perhaps we could learn more from the Northwestern University School of Medicine officials. The chief of Charlie's Hematology and Oncology division from 2006 to 2010 flat-out refused to talk with me. Odd again. Not long after being refused Northwestern University officials' cooperation, we received a letter from its official legal counsel stating a general refusal of any Northwestern employees to discuss anything about Charlie.

Were they hoping we would drop our interest? I was so surprised that, instead, we decided to follow the threads of the case that reminded me of my wife's mystery novels. What was going on here? The one Northwestern-related person who agreed to speak with me was Steve Rosen, formerly head of the Robert H. Lurie Comprehensive Cancer Clinic at Northwestern University. He had moved to Pasadena in 2013, where he was now provost and chief scientific officer of the City of Hope, the fifth-largest cancer center in the United States. Rosen told me that he believed that between 1994 and 2010 when Charlie left Northwestern, Charlie had "saved more lives than anyone in American medicine."[1] Wow! He also commented on Charlie's generosity in training young medical students and doctors. Those personal traits and follow-up actions granted him hero status among researchers.

If he had indeed saved many lives, why had he been chased out of Northwestern? A grant-accounting mistake? Surely not. Things were not adding up. In 2014 I called Linda Wyetzner, counsel for the accuser (an administrative assistant named Melissa Theis) in the charge of allegedly misusing grant money. Wyetzner was very friendly on the telephone and seemed certain of the strength of her case. "He committed fraud—that's what I had to prove under the Act." But when I probed for specific examples, she seemed either unwilling to provide or incapable of providing specific details.

Kurt Lindland, assistant US attorney for Northern Illinois who was largely responsible for the decision by his office to support the case against Charlie, answered my calls. He was gruff and certain. He told me that Charlie, as principal investigator of major federal government grants, had approved payment requests that were in fact from fraudulent entities. I asked if he could establish that Charlie was aware of the fraudulent nature of the claims. No, but Lindland insisted that "he was in on it." He did not offer any proof. When I asked if he was aware of and took into account Charlie's medical research achievements, he assured me that he was focused only on whether Charlie had violated the law by "abusing" grants. "This office has been willing to take on anyone." He rolled out the names of the rich and powerful that his office had doggedly investigated: Governors Jim Ryan and Rod Blagojevich; Congressman Jesse Jackson Jr.; Mayor Richard Daley; and former White House Chief of Staff Scooter Libby. It was obviously a matter of pride to him that status did not protect those who violated the law. Later on, it occurred to me that perhaps Charlie's eminence had worked against him.

Even some of the most heinous criminals get some credit for the good they have done: "He was an Eagle Scout," "He took care of his sister's puppy." Here, though, the DOJ tried to destroy Charlie both professionally and personally . . . for his hubris?

As we continued our research, we became increasingly convinced of Charlie's innocence and basic decency. Many factors played a role in my arriving at this conclusion. First was Charlie's willingness to answer any question we asked. And nothing that he told us has shown to be untrue. Second was the unanimous praise and admiration he received from former patients, collaborators, and aspiring clinicians and scientists whom he had mentored. Third was the weak nature of the documents meant to establish his guilt. Several of the people that we interviewed suggested that he was the victim of a plot hatched by pharmaceutical companies, whose profits he had threatened with his drug safety research.

We also became convinced that Charlie Bennett also was constantly motivated by the desire to achieve something scientifically important. He is, we came to believe, a good man, somewhat self-centered, with the soul of a progressive activist, happiest when he is busiest with activities that he believes will help people and enhance his status.

This book chronicles the professional life, research, successes, and struggles of Dr. Charlie Bennett. Some readers will find it a medical-practice mystery; others will discover that for-profit grants to academe secretly tarnish the ivory tower; others will learn that the highly revered Whistleblower Act of 1989 can be used to destroy individuals who stand in the way of big money and big profits.

Big Pharma is not an easy opponent. It is rich, powerful, and notably generous to its supporters. Its generosity has regularly benefited academia by way of significant grants to institutions and scholars addressing topics relevant to medicine and biology. Because of their generosity, both institutions and scholars are eager to win and keep the goodwill of Big Pharma. As we also learned, for those who differ from their interests, large drug companies can be formidable adversaries.

This book is told in the voices of the many people involved in Charlie's saga, which is a microcosm of the battle between Big Pharma's profits and scientific integrity. First, we introduce the reader to Charlie, a successful workaholic who created an international team to uncover adverse reactions to commonly prescribed drugs. We describe reactions of Big Pharma, academia, FDA, and the Department of Justice to Charlie's discoveries and the punitive consequences to Charlie Bennett. One of the first things we learned was that the successful pattern of his life was upended. Almost immediately after he publicly revealed the dangers of a blockbuster drug, he was sued by a university account clerk, accusing him of accounting fraud.

A major blow was an assistant US attorney for the Department of Justice joining the lawsuit; as a result of both the lawsuit and the widespread gossip it engendered, Charlie felt under constant attack and decided to leave his position at Northwestern.

In this book, we follow Charlie from Northwestern University to his new South Carolina University School of Pharmacy, where he demonstrated that, once again, other bestselling drugs produced serious or fatal side effects. We allow survivors of the blockbuster drugs Levaquin and Cipro to tell their stories and their appreciation for Charlie's battle on their behalf with Big Pharma and the FDA.

At the same time that Charlie was saving lives by uncovering the side effects of drugs, he was engulfed in an endless legal battle over the earlier fraud accusations. We show that, despite unreliable evidence, his accusers continued to insist he was guilty.

We conclude by examining three great institutions that play a dishonorable role in this story: the drug industry that produces miracles yet bitterly resists valid criticism; our universities that too often seek to stifle legitimate faculty research in pursuit of drug company support; and the justice system, which makes it difficult for the falsely accused to establish their innocence. Step off the cruise with us, examine the evidence, and decide where fault lies.

CHAPTER 1

Charlie Bennett and Adverse Drug Reactions

Steven T. Rosen, provost and chief scientific officer of the City of Hope National Medical Center in Duarte, California, concludes "Charlie Bennett saved more lives than anyone in American medicine by his major role in demonstrating the dangers created by the billions of dollars of erythropoietin medications, sold by the pharma giants Amgen and Ortho,"

Despite his credentials and experiences, Dr. Rosen said he has no idea how a drug designed to save lives had these consequences—and why Amgen and Ortho did not themselves reveal those consequences.

The day-to-day grind of practicing medicine bored Dr. Charles Bennett senseless. He was certified in internal medicine and also medical oncology, which generally involves seeing patients. He did work with the Veterans Administration patients, but "What I wanted was a blend of clinical oncology with health policy—a place where I would be creative." He listened to the vets; he studied patterns. He tested his own hypotheses.

Not many positions at the forefront of health policy were available for MDs, even board-certified oncologists; still, he hoped that, if he added a PhD, he could follow in the footsteps of his idol, childhood neighbor, physician, Nobel Laureate Professor Jonas Salk. He researched opportunities; quite uncharacteristically, he groveled before university administrators before the RAND Corporation accepted him in a PhD program. Things got exciting. In addition to the coursework and writing his dissertation, Charlie had been selected to treat HIV patients for the UCLA School of Medicine, attend medical and health policy seminars at RAND, UCLA, and the West Los Angeles Veterans Administration Medical Center, and help professors at RAND, UCLA, and the VA with their publications and other research. In 1989, he received his PhD in public policy with honors in Social Science Methodology.

Charlie was ready to set the medical world on fire. In 1991, now armed with both medical and policy degrees, he was appointed to his first academic position as assistant professor at Duke Medical School. He focused his writing and research on AIDS and was soon the recipient of major grants totaling over $2 million.[2] His prolific, original writing brought him to the attention of other medical schools. Feeling the sky was indeed the limit, three years later he was offered and accepted an appointment as associate professor at the highly prestigious Northwestern University School of Medicine as the A.C. Buehler Professor in Economics and Aging.

At Northwestern, his academic productivity and creativity continued. Between 1994 and 2010, he was either principal author or significant contributor in over 275 refereed journal articles and roughly fifty additional articles in non-refereed journals. His publications initially concentrated on various aspects of AIDS and then moved to developing cancer screening and treatment programs for inner-city persons with cancer in Chicago. These studies provided a pathway he could well have continued throughout his career.

But an unusual sequence of events altered the focus of his scholarship.

In 1997, his Northwestern Hospital colleagues presented a case of a patient at the hospital suffering from thrombotic thrombocytopenic purpura (TTP), a rare and potentially fatal syndrome. Her attending staff were treating her with Ticlid, a new, heavily advertised aspirin substitute; the FDA had approved that Roche substitute for heart patients in 1989 to decrease the risk of strokes and heart attacks. Charlie was instantly intrigued—TTP wasn't a run-of-the-mill illness.

Charlie decided to visit the patient in the Intensive Care Unit. When he first saw her, she was on a ventilator. He could barely make out her features. When he did, he realized with surprise that she was Jan Meister—a close friend of his father's. Even more interested in TTP now, he discussed her symptoms and treatment with her attending physicians, opening a new folder of articles he always carried. After their discussion, and within a few weeks, Jan Meister was off the ventilator and treated with plasmapheresis, the lifesaving blood-filtering treatment for TTP, and on her way to recovery.[3]

In the aftermath of this incident, Charlie wondered about the value of Ticlid: was it truly a wonder drug as Roache spokespeople claimed, or was it dangerous to heart and stroke patients?

Three months later, on the cross-channel train between England and France, Charlie was reading articles and scientific books and serendipitously struck up a conversation with yet another stranger—a fellow American, John Victor, who owned a California Hemacaret facility. By the time the train reached Paris, Charlie knew as much about John Victor as his old friends did. A few months later, Charlie, still wondering about the relationship between Ticlid and TTP, decided to visit the

Los Angeles plant of Victor's company. Of course he did. He met with the head nurse and asked if she could find out if any of their recent patients had been treated with Ticlid. She could. She told him that they had treated five patients with Ticlid two weeks before the patients developed TTP. As he listened, he flipped his tongue upside down and back, upside down and back—an early habit that helped his concentration. When he returned home, Charlie contacted CobeSpectra, the largest manufacturer of plasmapheresis equipment for treating TTP in the United States. Using a bit of charm and lots of technical language from his experiences, he talked to the VP for Operations into helping him contact the directors of the largest TTP treating centers in the country. Through phone calls to these medical directors, he uncovered twenty additional Ticlid/TTP cases. From a friend who previously worked in drug safety at the FDA, he located thirty-five more de-identified reports in FDA files for a total of sixty Ticlid/TTP patients.

He also discovered that twenty-two of those sixty patients had died from TTP.

Charlie published these dramatic findings for other physicians to consider in 1999 in the *Annals of Internal Medicine*.[4] Charlie's findings were important enough that the *Wall Street Journal* reported that sixty patients

> who received between two and six weeks of treatment with Ticlid and who subsequently developed the potentially fatal blood disease called thrombotic thrombocytopenic purpura, or TTP, between 1991 and 1998. Most of the patients who contracted TTP were taking Ticlid to prevent recurrence of stroke, and one-third of the TTP patients died from TTP.[5]

Charlie's human interests, and his outside-the-box approach to drug research, altered the development of that widely used drug. The resulting publicity gave him acclaim both at Northwestern and among medical researchers throughout the country. His increased notoriety made it far easier to obtain grants for his future work. Indeed, he received over $2 million just to study Plavix and potentially new cases of TTP related to Ticlid.[6] Charlie began to hear from other physicians and pharmacists about unexpected side effects of other drugs, their "adverse reactions."

This network of information motivated him to form a group of experts to study clinical experiences to first locate, and then deal with, adverse drug reactions. In 1998, Charlie's multidisciplinary team of investigators created RADAR (research on adverse drug events and reports), a clinically based post-marketing surveillance program. Their new program would systematically investigate and disseminate information describing serious and previously unrecognized adverse drug and device reactions (ADRs).

"Drug research" took on new meaning to pharmaceutical companies, their investors, and the public. Soon, the National Institute of Health provided million-dollar grants for RADAR; Charlie used some of that money to mentor young student scientists who worked on RADAR-related projects. In the process of investigating facts for this book, we met and talked with many of those scientists who are now in their own labs or teaching in universities.

Ten years after RADAR began, Charlie and nineteen coauthors announced their astonishing successes in the *Journal of the American Medical Association* (*JAMA*). "RADAR investigators identified sixteen types of serious ADRs among 1,699 patients, of whom 169 (10 percent) died as a result of the reaction."[7]

While working with RADAR and helping it obtain grants, Charlie continued to study adverse drug reactions on his own under his TTP grant. One of his first findings was that Plavix, the drug developed to replace Ticlid, also led to an increase in TTP. Doctors prescribed Plavix in the belief that it was safer than a pharmacologically related drug, ticlopidine or Ticlid. Ironic, but sad and potentially dangerous. As explained in the *New York Times,* April 21, 2000, "Dr. Charles L. Bennett's team of the Veterans Administration Healthcare System in Chicago has linked Plavix to thirteen cases of TTP." The *New England Journal of Medicine* posted the findings the next day on its website (www.nejm.org) to alert doctors of their study's potential importance for standard medical practices. Charlie's TTP grant had allowed him the time and resources to uncover dangers to the public. Although usually hesitant to change warnings, the Food and Drug Administration said it would change the warning label on Plavix in a hurried-up several weeks to account for the potential dangers of TTP—and they did.[8]

CHAPTER 2

Charlie and the EPO Revolution

Researchers typically work in their offices or labs, and they rarely receive the federal results that Charlie's discovery on adverse drug reactions did. The Blue Cross/Blue Shield Technology Assessment group added Charlie to its international team of researchers. He and five other oncologists and scientists created a research team to evaluate a new class of biotechnology drugs, called ESAs (erythropoietin stimulating agents). Charlie easily switched gears from TTP to ESAs. To enlarge their scope of doctors and patients, the assessment group allied themselves with the Cochrane Collaboration, a well-funded international enterprise that collects materials on scientific topics.[1] After fulfilling necessary protocols, they become a "Cochrane review group," which allowed them access to data sets already gathered by the Cochrane collaboration on ESAs. There was much to learn before they could set out their own protocols and elicit new data.

ESAs were the product of a major collaborative investigation that began in the 1980s, initially in response to the AIDS epidemic and the worldwide contamination of the blood supply with HIV. In the 1980s, a University of Chicago chemist, Eugene Goldwasser, proved that erythropoietin (EPO, a protein produced mostly by the kidney) stimulates production of red blood cells and is necessary for the body to produce hemoglobin.[2] More EPO means more red blood cells, which means organs like the heart and brain can function more efficiently.

In the mid-1980s, after Goldwasser had worked for decades on erythropoietin, he was enlisted as a consultant by a then-small pharmaceutical start-up company, Applied Molecular Genetics (Amgen). Goldwasser didn't have the resources to move forward; Amgen didn't have Goldwasser's understanding of EPO. They made a major medical-world coupling. Working together, Amgen scientists and Goldwasser developed a technique for producing EPO in large quantities through recombinant technology.[3] In 1987 Amgen was able to patent the underlying recombinant technology,[4] which it used to form a new EPO-stimulating substance, Epogen.[5] The ability to produce EPO on a large scale and its potential to meet worldwide needs to overcome anemia or increase energy were enormous. That patent was a valuable resource. Amgen could prohibit other companies from producing EPO, or it could sell the right to one of the companies eager to use the patented technology. In 1995, Amgen licensed the right to market EPO to cancer patients and to pre-dialysis patients for a large sum of money to Johnson & Johnson (J&J), which created a major new unit, Ortho Biologics, to market its EPO brand.

The Cochran research group had an easier time understanding all this than we did, but what we learned was that the first target for selling the stimulating agents, ESAs, was to kidney patients on dialysis.

Early tests showed that the ESAs, by increasing hemoglobin, substantially reduced the need for blood transfusions for dialysis patients. The tests did not uncover any adverse reaction—not that everything went smoothly. In the 1990s, the safety of ESAs came into question when multiple international doctors reported similar life-threatening reactions. The reports followed 281 patients on dialysis in the United Kingdom, France, Spain, Italy, Singapore, and Canada who had received EPO, but had then developed a serious, and previously rarely reported, syndrome: "pure red cell aplasia"[6] that had to be treated with blood transfusion every other day.

When Charlie and his group investigated these reactions, he sliced and diced and discovered the subcutaneous administration of J&J's EPO formulation was to blame. J&J had slightly altered the formulation in 1998 due to "mad cow disease" associated with the protein albumin. Most of us had, of course, heard of mad cow disease, but few prominent reports disseminated the slight manufacturing change that led to the terrible outcomes in those six countries. This international research group studied the widespread problem. They narrowed the tests, and Charlie wondered if, and hoped that, they could convince physicians to discontinue the subcutaneous administration specifically of J&J's EPO to dialysis patients.

The side effect disappeared by 2002. No antibodies developed. European medical groups outlawed subcutaneous administration in 2002, and all of the cases of pure red cell aplasia went away.As Charlie noted:[7]

> Because the mystery deaths were explained and because the injection-correction of adding a Teflon stopper to the EPO vial was minor (2006)—compared to taking the J&J formulation of EPO off the market—I was suddenly a pharmaceutical hero. J&J could restart marketing their EPO formula to persons with kidney disease who were on dialysis in the six countries of

interest; EPO sales for Johnson & Johnson in those six countries brought in several billion dollars annually.

It's easy to understand why, at this point in his career, Charlie Bennett had positive relations with both Amgen and J&J.

But then. Then he investigated the biasing effect of Big Pharma's sponsorship on research. In 1999, in collaboration with an Amgen economist and student assistants, he published in *JAMA*[8] a thorough examination of the existing scientific literature about pharmaceutical sponsorships. The article concluded that "pharmaceutical company sponsorship of economic analyses [of clinical trials] is associated with reduced likelihood of reporting unfavorable results." Thus, he concluded, if Big Pharma sponsored the trial, the results tended to be favorable to the drug; on the other hand, most of the published studies that concluded EPO *was not* cost-effective in the oncology setting were written by authors *not* funded by pharmaceutical company grants.

This was not an incidental or unimportant fact. It was a central part of the effort by both Amgen and J&J to win the allegiance of oncologists and hematologists throughout the country. Not only were those doctors given a variety of valuable equipment, but the salespeople were able to point them to a large array of companies-funded articles and studies. Charlie's surprising finding was funded by an unrestricted grant from Amgen, which, as you can imagine, was not Amgen's intention: "Charlie Bennett from Northwestern U has been on a safety crusade . . . he has offered his document for our critique."[9]

Charlie was not making friends in the fierce pharmaceutical circles where, with billions at stake, the competition for market share between Amgen and J&J became intense. Both sides stepped over ethical lines. They fought this battle in hospitals and doctors' offices, where prescriptions are conceived. They offered well-paid speaking engagements to prominent physicians whom they referred to as "Key Opinion Leaders," a term they often used in letters of solicitation.[10] They also offered physicians products at greatly reduced prices, and they hired and trained platoons of sales representatives, whom they designated in terms meant to suggest professional medical skills such as "medical consultants." In some instances, a physician who prescribed high-dollar amounts of EPO could receive an additional several hundred thousand dollars for working in his or her spare time with Amgen.[11] The underwriting of physicians was an effective form of advertising and of acquiring allies.

Enthused by EPO's early success in almost eliminating the use of blood transfusions for persons on dialysis and for persons with cancer, scientists began to speculate whether increasing hemoglobin levels above recommended levels in the package insert would not only decrease the need for red blood cell transfusions but also fight cancer. It was a heady fantasy.

One of the first to test the increased-hemoglobin level thesis was German scientist/radiation oncologist, Michael Henke, who was already in the fight against cancer. He initiated a randomized clinical trial (ENHANCE) using the Roche Pharmaceutical version of EPO. Henke measured the survival effects of large doses of Roche's EPO to head and neck cancer patients who were receiving radiation therapy. The results surprised and dismayed him. He uncovered a 1.27-fold increased risk of death among cancer patients who received EPO. No one had anticipated that. Henke's 2003 summary report, published in *The Lancet*, was terse and evocative: "Epoetin beta corrects anaemia but does not improve cancer control or survival."[12]

He urged other researchers to conduct follow-up studies of EPO's safety.

Amgen and J&J did not publicize Henke's distressing research results; indeed, the negative results might have been ignored altogether if Henke's findings hadn't acquired a significant champion in Paul Goldberg, the editor-in-chief of *The Cancer Letter*.[13] Goldberg, a tough-talking, quick-thinking medical reporter from New York via Moscow, played an important role in bringing out relevant information about safety concerns with EPOs. Goldberg realized Henke's EPO findings were important and began to use Henke as a resource for *Cancer Letter* articles.

Goldberg also knew and admired Charlie's work on cancer and EPO, and from time to time would contact him for comment or insight about a story he was working on. Goldberg thought Henke and Charlie were similar in important ways—both constantly working, and, importantly, both willing to challenge generally accepted truths. At Goldberg's suggestion, Charlie looked at the 2003 *Lancet* article from Henke's group describing the origin, methodology, and results of Henke's clinical trial.

Charlie was favorably impressed with both the study and the courage of Henke, and he was stirred by Henke's urgent call for follow-up research. Charlie and Henke met at the 2004 American Society of Clinical Oncology (ASCO). They hit it off—much as Goldberg predicted.

Henke convinced Charlie that physicians worldwide were making a mistake by their failure to follow up on his findings of the safety and mortality problems with EPO administration to cancer patients.

Henke also alerted Charlie to the possible cause of the higher death rates among head and neck cancer patients who received EPO and radiation therapy—cancer cells in these patients might contain erythropoietin receptors that would provide a base for cancer cell growth.[14] He theorized that when EPO was administered during the time that radiation therapy was simultaneously being administered, the EPO had the exact opposite effect the designers and manufacturers hoped for, i.e., causing cancers *to grow* rather than to get smaller. Amgen scientists still challenge this theory today.

Charlie came away from their meeting with two thoughts: first, that he needed to study the possible negative effects of ESAs, and second, that he would like to work with Henke.

Around the same time, another prominent researcher/breast cancer specialist, Bryan Leyland-Jones, also strongly believed in the higher-hemoglobin theory. He held a multicountry clinical trial of the effect of EPO on outcomes among women who were treating breast cancer with chemotherapy; optimistically, he expected to find that more EPO-induced rises in hemoglobin levels brought more blood to the tumor and allowed chemotherapy to kill more cancer cells. What a disappointment . . . and he too had to terminate his study due to high mortality rates.[15] Nevertheless, at a scientific conference convened by the National Cancer Institute, Leyland-Jones continued to argue that it was still possible that EPO could be valuable.

Later, when Leyland-Jones recognized the limits of EPOs, he described his study's failure and his fluctuating conclusions about EPOs in his fascinating article "Erythropoiesis Stimulating Agents: A Personal Journey."[16] There, he offered contrition and sadness concerning his earlier "strong anticipation of a strong positive ESA survival effect." He admitted that "the concept of ESAs of pushing hemoglobin levels to 14g/dL (20 percent greater than the normal level) was wrong."

In 2006, an ongoing Danish study, Dahanca 10, conducted by head and neck surgeon Jens Overgaard, cast further doubt on the value of EPOs. The study was carefully constructed,[17] focusing on Amgen's EPO product "Aranesp." Once again, researchers had to terminate the study early because interim data analysis revealed that the cancer group receiving both chemotherapy and Aranesp did significantly worse than the control group. Goldberg's *Cancer Letter* publicized the interim results in the United States; the results also appeared on Google.[18] Not surprisingly, Amgen took strong exception to the Dahanca 10 findings, to *The Cancer Letter*, and personally to Paul Goldberg, who had refused to answer an Amgen subpoena concerning the Dahanca study.[19] The judge upheld Goldberg's right to refuse under the First Amendment, "As a member of the press and as a naturalized American, I consider it a privilege to defend our First Amendment rights against attacks from one of the world's largest biotechnology companies."

As a result of these aborted studies, the Cochrane Collaboration's Overview Report on EPO administration to cancer patients was more cautious. Authored by Julia Bohlius, Charlie, and Ben Djulbecovic, the report acknowledged a potential harmful aspect of EPO treatment in the cancer setting—but they did not report on the Henke, Leyland-Jones, and Danhanca 10 studies. They left the door open for further study—and doubts. They wrote that EPO and DARB "reduce transfusions in aneamic cancer patients," but, they concluded, the drugs also increase the risk for blood clots. "Whether EPO/DARB affects survival is uncertain. Uncertainties remain as to whether and how EPO or DARB affects overall survival."[20]

In January 2006, Charlie received an invitation to be the plenary speaker at the Society of Clinical Trialists (OSCT) meeting in Montreal. He prepared a paper that dealt with EPO and DARB and the increased risks of blood clots with these drugs. Just before he was to speak, Charlie received an excited phone call from his collaborator Ben Djulbecovic, who said that *any* amount of EPO or DARB could actually increase the chances of blood clots and death: he had analyzed the new Cochrane data. They showed a significantly higher mortality rate and blood clots *with* EPOs or DARBs than *without* EPOs or DARBs, across the board.[21] The risks were no longer "uncertain."

Charlie quickly added new slides that did not track the content of their original paper but instead outlined the association of EPO/DARB use and increased mortality in the cancer setting. Several Amgen PhD scientists and statisticians were in the audience in Montreal, and one of them sharply questioned the statistical methodology.

Subsequently, Charlie, Bohlius, and Djulbecovic coauthored an abstract submitted for the 2007 annual conference of the American Society of Clinical Oncology (ASCO) in Chicago. The abstract reviewed the Cochrane group's findings of increased risks of blood clots with EPO/DARB administration to cancer patients.[22] Unfortunately, the once-congenial group was soon to split. Charlie urged that the coauthors add Djulbecovic's new mortality findings to the presentation in

Chicago. Bohlius was uncomfortable with the conclusion, which flew in the face of their earlier work. They had a major dispute about the solidity of the findings. Charlie, now confident in the analysis, insisted on including the important mortality data. There was no compromise available, and Bohlius demanded Charlie remove her name from the list of coauthors for this presentation. Totally believing in Cochrane's earlier work and distressed with Charlie, she proceeded to have him removed from the ongoing Cochrane Collaboration and unsuccessfully attempted to have Djulbegovic removed as well.

The American Society of Clinical Oncology (ASCO) did not allot Charlie's Chicago abstract a highlighted, oral podium presentation. Instead, he was to present his findings at a poster presentation—in collaboration with an undergraduate Northwestern University research assistant. They learned that their non-highlighted presentation would be allocated to the far rear of the large hall, where it was likely that a disappointingly small number of participants might drop by to learn about this new correlation related to blood clots and mortality . . . important findings and controversies were rarely relegated to a back corner.

Charlie had sent a draft of his proposed paper to Goldberg, who thought the new data was important: "According to Bennett's analysis of data from randomized trials reported since 2003, patients taking these agents face a 59 percent greater risk of VTE and an 11 percent increase in the risk of death, compared to patients who don't take ESAs." *The Cancer Letter* took Charlie's research and resulting questions seriously:

> "We have a safety signal on deep vein thrombosis, and we now have a safety signal that we see on survival," Bennett said to *The Cancer Letter*. "How much more do we need to show you to stop overuse of these drugs? How many safety signals do we need before we get to the idea that we have to reconsider what we are doing here?"[23]

Goldberg's urging led to a large crowd, filling the poster presentation's allotted space and more.

By the time of the presentation, Charlie was less sure of his conclusions and worried that he was taking too strong a position with respect to EPOs; Bohlius, whose abilities he had always respected, had left his project because of disagreement with the mortality findings.

Additionally, the updated data had not even been accepted for publication—the *Journal of Clinical Oncology* rejected his paper about the substance of his Montreal presentation. In his submission, Charlie pointed out that studies before 2003 focused on the effect of ESAs on hemoglobin levels; after Henke's studies, they focused primarily on mortality instead. For instance, Charlie learned that the *Journal of Clinical Oncology* editors felt that Charlie introduced an artificial before-and-after 2003 "cut-point," which made his conclusions unreliable. That cut-point became the focus of the next years of Charlie's research.

The day before Charlie's ASCO presentation, Charlie and Djulbecovic met with Amgen's then vice president Roy Baynes[24] in the lounge of the hotel. Charlie had sent Amgen a draft of his presentation.[25] Baynes insisted that the paper was unworthy of publication.

As Charlie recalls, Baynes ended his criticism with an ominous threat in front of coauthor Djulbecovic: "If you publish this, I will destroy you."

Charlie reports that many Amgen employees and advisors accosted him and questioned his findings and methodology.[26] He even heard by unsubstantiated rumor that Amgen had induced several important scientists to criticize his conclusion—including a Nobel Laureate.

Thus, prior to the Sunday poster presentation, Charlie was nervous, fearful of retribution from Amgen, and worried about how the audience would respond in light of the presentation's heavy Amgen presence. Would Amgen, led by Baynes, really try to destroy his career? Would he be viewed as a charlatan?

For once, his self-confidence wavered. His career was in danger. His tongue flipped back and forth much of the night. At 4:00 a.m. Sunday, he called his Northwestern friend Steve Rosen. Charlie was unusually inarticulate, which helped to convey his anxiety to Rosen. Rosen urged him to stick with his conclusions. However, when Charlie asked if Northwestern University would back him up against charges from Amgen, Charlie recalls that Rosen said, "We are not in a position to back you up. It's your work. Neither I nor anyone else here has studied it. You are the one who has to stand behind it." Perfectly sound analysis—but not advice likely to calm apprehension.

Four hours later, a tired and uncomfortable Charlie looked around at large crowds with eager reporters stuffed into his small poster spot; his fears of making a costly mistake grew stronger.

He gave into his fears.

He began by announcing that "We break no new ground." He paused and then announced that nothing in his own recent findings indicated new, unannounced danger signals when patients did not receive EPO or DARB targeted to high hemoglobin levels. Thus, because Charlie now reported that they had observed no new safety signals in their updated analysis, the newsworthiness of his poster presentation was lessened—markedly.

Paul Goldberg was angry. His friend Charlie was standing right in front of a poster that said just the opposite. Instead of writing the supportive *Cancer Letter* article he had anticipated writing, he wrote one expressing his sharp disappointment:

> CHICAGO—Last week, Charles Bennett prepared to present a provocative and potentially important finding: According to his meta-analysis, patients who took erythropoiesis-stimulating agents had a statistically significant increase in the relative risk of venous thromboembolism and death. But instead of delivering a message of caution, Bennett said that his findings weren't newsworthy, and that ESAs pose no previously undisclosed risks to 90 percent of patients.

In private, Goldberg confronted Charlie with a single phrase to reflect his profound disappointment: "What the fuck, Charlie?" Goldberg was angry, but the media's take on the presentation included Roy Baynes, who accepted the middle ground for a restricted use for anemia cancer and drugs: "certain patients, such as those with head and neck cancer where the doctors are trying to drive hemoglobin levels higher, should not use the drug off-label."[27]

Unhappy with Goldberg's scorn and uncomfortable with his own cautious retreat, Charlie responded as he always did to challenges—by doing more work. Returning to Northwestern, he and Djulbegovic worked rapidly and intently. Were mortality risks greater when EPO and DARB were administered to achieve lower hemoglobin targets? They had not focused on that statistic before the 2007 Chicago presentation, and now they did.

He enlarged and reconstructed the study analysis. New data became available, and they found that several reporting clinical trials pointed to the dangers of ESAs that were independent of the target homoglobin-level data. For instance, the *Journal of Clinical Oncology*[28] reported on a Canadian study of lung cancer patients receiving EPO. Oncologists there had initially designed the study to investigate the "quality of life effects of EPO," and they were not then concerned with targeted high hemoglobin levels. Yet, once again—as was true of the earlier studies from Denmark, Germany, and the United States with cancer patients receiving EPO or DARB—the researchers had to abort the Canadian study because of mortality concerns among cancer patients who had received EPO.

Charlie, in collaboration with Djulbecovic and Michael Henke, produced a new, more powerful manuscript that *JAMA* accepted for publication.[29] *JAMA* released this important article just three weeks ahead of a previously scheduled March 2008 meeting of an FDA Advisory Committee. Now Charlie's research team had a high-level professional, peer-reviewed journal's acceptance, their statistics, and their conclusions. Charlie felt a new surge of confidence. The FDA group specifically designed this meeting to review the safety of EPO and DARB when administered to anemic persons with cancer or with kidney disease.

The *JAMA* article concluded that EPO increased the risk of death of cancer patients by roughly 10 percent—and that the mortality risk was *independent* of the target hemoglobin. This painstaking study, by its very size (fifty-one clinical trials with more than 13,000 cancer patients), permitted scientists to come to conclusions about the mortality risks of EPO and DARB that were not possible previously. Andrew Pollack, the *New York Times* writer who had covered Amgen and J&J extensively, warned of

> a statistically significant increase in the risk of death from the drugs. . . . Widely used anemia drugs sold by Amgen and Johnson & Johnson raise the risk of death among cancer patients by about 10 percent, according to Dr. Charlie Bennett's new analysis of previous clinical trials that was published Wednesday.[30]

He pointed to Charlie's "new ground": Charlie said that "he did not believe the higher risk of death came from those blood clots." Rather, he said, "there is evidence that the drugs, which are synthetic forms of a natural hormone called erythropoietin, directly stimulate the growth and spread of tumors."

Pollack's story announced the startling findings to general readers. So, too, did Goldberg announce them to the science community. "When you put all these studies together, there is a 10 percent increase in relative risk of death and a 57 percent relative risk of VTE," said Charles Bennett.

Less than six weeks after the group's research findings were made public in *JAMA*, the FDA Oncological Drug Advisory Committee met to discuss increasing the regulation of EPO and DARB. By the time of the meeting, the evidence had become too powerful to ignore or summarily dismiss. Patients, experts, and lobbyists all made passionate arguments.

Probably the most dramatic moment of the meeting came from Dr. Otis Brawley, a member of the advisory committee. A longtime friend of Charlie's and a well-respected oncologist scientist at the National Cancer Institute, he, too, had come to believe that ESAs increase danger for patients. He concluded that he needed to challenge the efficacy of EPO drugs. As he reports in his autobiographical book *How We Do Harm*, cowritten with Paul Goldberg, "It's up to me—to make the press and public understand that this is not an obscure scientific controversy."[31]

During the advisory committee discussion, Dr. Brawley summarized the data and posed a question to those supporting continued acceptance of EPO-stimulating agents, ESAs: "I am concerned that this compound is a 'tumor fertilizer' . . . What data do you have to assure me that this is not Miracle–Gro for cancer?" No effective response.

To his satisfaction, Brawley's quip had its desired effect; the *New York Times* quoted his question on the front page, as did almost every news story covering the meeting.[32]

The committee advised the FDA to request that EPO and DARB manufacturers add safety information to the product labels, indicating that those drugs should not be administered to anemic cancer patients who were receiving potentially curative chemotherapy treatments. The committee overwhelmingly voted that the FDA should issue a ruling that required physicians were no longer allowed to expose chemotherapy patients to EPO or DARB—because of the increased risks of death.

However, it was not until 2010 that the FDA followed its own committee recommendations. They finally instituted a Risk Evaluation and Mitigation Strategy (REMS) that formally requires patients, pharmacists, and physicians to sign informed consent that they are aware of the mortality and blood-clot risks when EPO and DARB are administered to cancer patients. Henke's theory and Charlie's team research each played a major role in the FDA's decision to limit the use of EPO and DARB in the cancer setting.[33]

Amgen continued to argue that the scientific positions advanced by Henke, Bennett, and Goldberg were erroneous; nevertheless, sales of EPO and DARB in the cancer setting fell by 95 percent during the next decade. The *Washington Post* reported the decline, or demise:

> For years, a trio of anemia drugs known as Epogen, Procrit, and Aranesp ranked among the best-selling prescription drugs in the United States, generating more than $8 billion a year for two companies: AMGEN and Johnson & Johnson. Even compared with other pharmaceutical successes, they were superstars. For several years, Epogen ranked as the single costliest medicine under Medicare: U.S. taxpayers put up as much as $3 billion a year for the drugs. Nevertheless, the status of Aranesp and its fellow ESAs as a super-selling wonder drug was over.[34]

By 2008, after the limits of EPO and DARB work had been exposed, Charlie Bennett's career was soaring. He was internationally recognized as a leading researcher on adverse drug reactions. Other scientists regularly cited his work; graduate students vied for the opportunity to work with him. The American Cancer Society's National Research Council chose him for membership. He was in demand as an international speaker and consultant at major American medical universities.

Charlie was delighted by the widespread acclaim and proud of himself for overcoming his fear of Amgen's retribution. Charlie's professional status seemed secure. During the eleven-year period between 1997 when he joined the Northwestern faculty and 2008, he uncovered fatal, or near-fatal, side effects of twenty widely prescribed drugs. He organized RADAR to enlist physicians in monitoring and evaluating the potentially fatal side effects of drugs. He developed a system for mentoring promising college graduates who intended to apply to medical school.

While achieving all this recognition, Charlie simultaneously contributed to and published hundreds of scholarly articles on a broad array of subjects, almost all concerning important issues of medical public policy.

This is the creative career Charlie sought, where he researched inside and outside the accepted paradigms to discover ways to help the medical community. As it soon turned out, however, his status was vulnerable. His discoveries and publicity led to a major, negative shift in Charlie's relationship with leading pharmaceutical companies.

CHAPTER 3

Charlie Comes Under Suspicion

Charlie returned from his speaking schedule, grabbed his suitcase, took a cab from O'Hare, and went straight to work. He went through his conference area and into his office, wading through all the mail and his own leftover messes. He was happy; *JAMA* had published his article on EPO, DARB, and their mortality rates; he had sent copies to everyone on his lists, and was ready to bask in the anticipated academic acclaim.

He joined other members of his RADAR team scattered throughout the two adjoining offices, getting caught up and stuffing their recent drafts into his suitcase. He called a cab and finally went home. As he greeted his family, they looked down and saw, again, what wasn't there:

his suitcase. "Oh, yeah . . . but it's not at O'Hare this time—it's in my office." He plopped on his bed, content.

It was the last fully content night of his life for the next decade.

Charlie's troubles seem to have begun in September of 2008, when Alice Camacho,[1] his account clerk, began challenging the legitimacy of Charlie's grant expenditures.[2] Outraged, Charlie sent an email to Steve Rosen:

> Alice had reported to her co-workers that she felt that my subcontractors were not performing high quality work and that she was determined to find fraud in the work of my subcontractors. The co-workers informed me that she called several subcontractors at home and requested information on their prior and current work activities. She then prepared a "Power Point" presentation identifying the suboptimal work of my subcontractors. I heard rumors in the hallways that she announced to her co-workers that she would "bring Dr. Bennett down."

Rosen was sympathetic, but he did not have jurisdiction over the chief of Hematology and Oncology Division who had, in fact authorized Camacho to study Charlie's accounts and to create a PowerPoint presentation to illustrate her findings.

The Northwestern staff and officials treated him like a suspect accused of academic misconduct. His chief criticized Charlie's agreements with his consultants and student mentees, many chosen from outside Northwestern. He was particularly displeased that Charlie did not hold a competition before he picked subcontractors, and, indeed, picked them from outside of Northwestern. He added additional criticisms of guidelines ignored and listed procedures violated. In sum, the administrators of Charlie's division worried that Charlie made up his own rules and ignored University requirements in handling his grants.

There was something to that.

They were critical not only of Charlie's slapdash approach to guidelines, but several raised the specter of a whistleblower lawsuit based on Charlie's grant filings that they considered vague and inaccurate.[3] Slapdash took on epic proportions. After a confrontational meeting with administration and staff critics, Charlie reported to Rosen:

> The chief joined staff members Alice and Angela Youngfountain,[4] where they all agreed that my behavior provided the basis for a whistle-blower suit, and that the sub-contracts written by Feyi [former assistant] appeared to indicate that these individuals had not actually performed any substantive work.

Whistleblower lawsuits are governed by the False Claims Act (FCA), which authorizes actions against those who claim a right to government money or resources "knowing such claim to be false, fictitious, or fraudulent." Charlie was furious and could not understand how high-level administrators and employees at Northwestern had concluded that his claims for routine reimbursement were "knowingly false, fictitious, or fraudulent." And if a high-level university official believed that a tenured faculty member was pushing the boundaries of legality, it made more sense for those officials to offer Charlie legal help to forestall future problems. No administrator sought Charlie's version of the accusations before the meeting and threats. No administrator walked him through remediating steps.

Charlie was bringing in millions of dollars in grant money annually. Perhaps those at Northwestern who were quick to accuse Charlie found his "I'm important" personality grating. But self-praise and importance are a standard currency in academic discussion; most division chiefs are able to deal with academic personalities in less-confrontational styles.

After being accused of violating the False Claims Act, Charlie was both angry and frightened. He called his former assistant, Feyifunmi Sangoleye (Feyi), to discuss Comacho's accusations. She told him that the accusations were part of a plot to force him out of the medical faculty. She also suspected that the chief had appointed Alice Camacho to work on his accounts for that very purpose. Charlie was stunned, then confused. He hadn't encountered any problems during the time that Feyi was his chief accounting clerk. Was his department chief actually trying to take him off the medical faculty?

Amy Bennett, Charlie's wife, recalls getting a disturbing warning from another worried faculty wife during a new intern/resident dinner: "The new chief was brought in to fire your husband and my husband." Amy recalls another example of unfriendly behavior: "The other thing I thought was strange: Charlie got his endowed chair at Northwestern's business and medical schools; they had a *big* celebration for him at Northwestern, right after he got his chair. This would have been in December 2006, right after Charlie's prostate cancer surgery. His Chief didn't show up, and I thought that was very strange. Why wouldn't Charlie's immediate superior show up?"

Charlie had little doubt about the ultimate source of his troubles. Amgen was behind it all. "Had to be." Charlie's suspicions have never lessened.

Throughout 2008, administrative inquiries continued through a series of email questions, focusing on Charlie's accounting practices. Their scope was narrow, but their inference pointed; the questioning could have ended in a few minutes. So, indeed, as we read through the emails and held interviews, we understood that perhaps Charlie wasn't paranoid in thinking that Amgen had to be involved in his troubles.

His department staff and officials continued to probe with further questions, this time narrowing them to the bona fides of consultants, including Professor Sara Knight, now a professor of medicine at the University of Utah and former deputy director for Health Services Research and Development Service of the Department of Veterans Affairs.[5] They also interviewed Dr. Elizabeth Calhoun, who was incensed at the questions (see p.90 and note 23); they investigated a major contractor (ATSDATA), and made spreadsheets of his paperwork inaccuracies. Charlie was furious. He again wrote to his friend Steve Rosen, complaining about the continued questioning.

> Angela and this new hire [probably Melissa Theis, a hospital accountant] have decided to micromanage the grant, disparage my reputation with the rest of my staff, harass my consultants, and raise allegations of financial and scientific misconduct. They have never read any of the work products from the grant. This approach is intolerable.

Rosen was sympathetic, but the most he could do was to award Charlie with a $30,000 personal fund grant for reimbursement for incidental expenses that occurred in the course of his scientific investigations—expenses unrelated to Charlie's NIH support and not pertinent to the investigation.

While Rosen's grant eased the pressure on Charlie with respect to small expenditures that occurred in the course of his work, it did not alter the fact a hostile administrative assistant with no scientific training was constantly questioning the millions of dollars in government grants—for which Charlie was the principal investigator.

Perhaps someone higher up in the administration was fostering doubts about him. Rumors swirled. Journalist Paul Goldberg heard rumors that Amgen hired a private detective to investigate and spread stories of malfeasance by Charlie.

The administration, in a wise tactical move, asked Rosen to join the questioning and to ask Charlie to explain, in detail, his choice of various consultants and subcontractors. One of the grant's largest expenditures was to ATSDATA from the University of Chicago at Illinois; Charlie initially approved a series of payments amounting to roughly $90,000. At the time he signed the invoices, his former clerk Feyi had explained that ATSDATA was a research arm of the University of Illinois at

Chicago. Backtracking through his files in response to the ATSDATA questions, Charlie found the approval slips for the money he approved to the University of Illinois at Chicago's survey research work—which he assumed was the ATSDATA. He thus felt secure in responding, "I have reviewed their invoices. These individuals have provided top-flight epidemiologic support for RADAR, TTP, and my prostate work. They have published articles in the NEJM that build on their epidemiologic work."

In addition to ATSDATA, Charlie was asked about his RADAR web designer, Earl Ripling.

Charlie did not have to research his answer: "Earl Ripling and Associates have led my efforts to improve my dissemination efforts. He has published extensively in *Reader's Digest* and the *New York Times*, and his work has included a long history of interactions with the pharmaceutical industry." Charlie included these additional facts—but did not mention the widely known fact that Ripling was a family member, indeed, his cousin.

They also asked for documentation for consultant Ben Djulbecovich, with whom he worked on his *JAMA* article on EPO. Charlie attached Djulbecovich's extensive CV to his return email.

His department immediately returned his emails with additional questions about ATSDATA—in bold: "what were the specifics of services" . . . "how are those services connected to RADAR" . . . "provide the CVs/background of all those listed as paid" . . . "give examples of publications or articles." The detailed questions also focused on Ripling's website production: provide a copy of his professional brochure and explain how this website connected to a scientific aim of RADAR's grants. One email criticized: "Making a connection to obtain more grant funding does not sound like a fulfillment of a scientific aim."

The invoice interrogation lasted all fall. In November, they asked for more documentation on Ripling's website, MedRed (informatics software for healthcare providers and patients), ATSDATA; they even required a CV for Dr. Elizabeth Calhoun, one of Charlie's mentees/consultants.

Reviewing this series of emails, we were constantly struck by their timing; these are not typical academic questions based on an assumption of good faith. They are the type of questions a lawyer might ask to prove fraud or deceit. Charlie was suddenly faced with this Columbo-like, second-look questioning. Exasperated, Charlie finally responded that he was "willing to go another round . . . when do we move on?" His research and scholarship were now entangled in the web of university accusations.

Naturally Charlie immediately connected the university's animus to Big Pharma's hostility to him, beginning with Amgen's vice president Baynes's earlier threat at the ASCO meeting to "destroy him." His suspicions were solidified by material he obtained when he agreed to testify against Amgen in a major, multistate lawsuit. An assistant US attorney for the Western District of Washington state, Peter Winn, was spearheading the case against Amgen, based on its efforts to both mislead consumers and to distort scholarly writing dealing with its products.[6] Winn sent Charlie a large box of relevant material that the court required Amgen to turn over for the lawsuit. Charlie was dismayed to find himself discussed in Amgen internal communications. Several Amgen officials stated that "something needed to be done" to limit Charlie's influence. (See chapter 10.)

Amgen even worried about Congress. On March 6, 2008, David Brier, senior vice president for global affairs of Amgen, sent an email to other top officials expressing concern about a letter to the FDA dealing with ESAs (EPOs) cowritten by Congressman John Dingell, chairman of the House Energy and Commerce Committee.[7] Dingell's letter, relying on the "dramatic findings" of Charlie's *JAMA* article, was highly critical of Amgen's marketing techniques for EPOs; VP Brier summarized, "Our challenge is determining how to effectively respond to the conclusions."

Charlie was shocked by the Amgen emails. As he reported to me, he explained all this to Amy, and "I cried." He knew now that the most aggressive force in Big Pharma considered him as an enemy to be attacked. And he felt certain that he had uncovered evidence of why Northwestern officials had turned against him.

CHAPTER 4

Charlie Sued, Lawyers Multiply

Although Charlie Bennett suspected his problems originated with Big Pharma, his immediate problems came from Northwestern employees. Key among them was Melissa Theis, a new purchasing coordinator for university accounts. Theis was a friend of Alice Camacho and shared her suspicions of Charlie's abuse of grant money. In November 2008, Theis resigned from her Northwestern position. Theis then hired Linda Wyetzner, a lawyer who specialized in false claims litigation.

Before they filed the suit, Theis contacted the FBI to present her suspicions about Charlie. According to the report of her FBI interview, Theis focused on suspicious expenditures of grant money.

Wyetzner investigated briefly and then, in March of 2009, filed suit under the False Claims Act (FCA), on behalf of Theis—naming as defendants Charlie and Steve Rosen and Northwestern University on the ground that they "failed to adequately supervise"[1] Charlie's expenditures. Although the lawsuit did not claim that Theis had been harmed in any way by the claimed violations, the FCA specifically permits a private person, known as a *relator*, to bring a lawsuit on behalf of the government. Such actions are known as qui tam proceedings.

The FCA required Wyetzner's complaint to be filed with the Justice Department. Afterward, the US Attorney's Office for the Northern District of Illinois investigated the allegations to determine whether the Department of Justice should join the suit against Bennett. Under the FCA, if the government intervenes, it and the relator become partners for purposes of the case. If they are successful, the whistleblower, or relator, is entitled to 15 percent to 25 percent of the amounts collected by the government.

The sequence of events from PowerPoint questions about routine expenditures to a formal lawsuit took little time but caused much damage for a long time.

As required by law, the complaint against Charlie that was filed in Chicago was sealed while the Department of Justice decided whether to join. Although the False Claims Act seems to contemplate a quick (sixty days) decision by the Justice Department, courts generally are willing to extend the period to permit a careful investigation. The qui tam against Charlie was under seal from 2009 until 2013. Thus, for four long years, neither Charlie nor his attorney had access to the complaint.

Wyetzner's complaint is remarkably devoid of specific allegations. We read and reread the complaint. Nowhere in the complaint is there any allegation that Charlie Bennett engaged in scientific misconduct; nowhere in the complaint is there any listing of actions that he took to enrich himself through fraudulent representations. Instead, the complaint attempts to equate violation of government accounting circulars and regulations with making a false claim. Thus, paragraph 13 states that:

> Defendant has violated, and continues to violate, the NIH Grants Policy Statement, OBM Circulars, and federal regulations that apply to: (1) basic cost principles; (2) the need to obtain prior approval for significant budget modifications; and (3) time/effort reporting procedures.

The US Attorney's Office for Northern Illinois asked Northwestern's University Counsel to make sure that no one destroy any items related to Charlie's grants while the investigation proceeded.

On February 1st of 2009, Northwestern University counsel, Amy Mayber, issued a Document Preservation Notice that was widely distributed to officials and physicians involved—in any way—in Charlie's research. Mayber instructed those who received the notice to preserve, with no changes, all documents related to any of Charlie's grants.

Charlie immediately contacted Mayber. She told him that, as he suspected, he needed a lawyer and added that the university would pay his legal fees for issues growing out of the investigation; she made clear, however, that neither she nor other university counsel would act on his behalf or consult with him about the investigation. They would pay, but not be a part of the suit, which gave the university the ability to later settle separately.

Amy Mayber strongly recommended that Charlie hire local attorney David Stetler to represent him; Charlie did so—unaware that Stetler's colleague David Rosenbloom was leading the university's defense team. Charlie later learned that Rosenbloom was simultaneously lead defense counsel for Amgen in the billion-dollar false-claims suit led by Peter Winn—the very suit in which Charlie had previously agreed to be an expert witness.[2] Thus, Rosenbloom was simultaneously representing Amgen in Washington, plus Northwestern and Charlie in Chicago.

Charlie often went to Stetler's office but did not learn much. Stetler limited himself to warning Charlie of the precariousness of his situation and the danger of a criminal indictment. Occasionally Stetler and Charlie joined Rosenbloom, who urged Charlie not to discuss—with anyone—the progress of the investigation. It all increased Charlie's already considerable apprehension.[3]

In the spring of 2009, Charlie Bennett gave a series of invited academic lectures on the West Coast. When he returned from his trip, he met in his office with Dennis West, one of the collaborators on his RADAR grant. While they were catching up on each other's news and activities, West pointed out that the office was its typical mess and suggested cleaning it up. Charlie looked around—nothing unusual. Most of the mess consisted of drafts of already finished and published articles. Still, West was probably right—so he dumped most of them into the wastebasket.

He didn't think about it until the next day when Douglas Vaughn, chair of the Medicine Department, informed Charlie that he was no longer allowed to enter his own office. Shortly thereafter, Charlie discovered that the lock on his office door had been changed.

It took several days of asking and calling and asking again, but finally Charlie learned through a series of emails that he had been permanently locked out of his office for violating the "preserve documents" instruction from the US Attorney's Office. Charlie quickly called on his lawyer Stetler, pointing out that he and Dennis West did not in fact dispose of "documents." He explained that he had merely shifted no-longer-needed drafts to his wastebasket. He reiterated that the drafts were totally unrelated to grants or any other expenditures.[4] Who had investigated his trash can?

Stetler refused to take any action about Vaughn's hostile behavior toward Charlie, stating that he had been hired only to represent Charlie in what he stated appeared to be an investigation that would end up with a criminal indictment of Charlie. As we read documents and talked to those involved, we ran across nothing stranger than the next episode in this saga. On May 22, 2009, Stetler made a confusing attempt to discover indirectly whether his client was involved in the criminal fraud. He requested that Charlie come to his downtown office immediately after completing his workday.

When Charlie arrived, Stetler told him that they were about to go into the law firm's meeting room, and that "this may be the most important meeting in your life. You will be asked a question, and your facial expressions will be observed by people important to the investigation and what they observe will be crucial to your future." When they entered the meeting room, Charlie observed eight men, whom he believes were attorneys for Northwestern, seated around a table. The group included David Rosenbloom.

As Charlie recalls the incident, as soon as he was seated, one of the attorneys asked him: "Did you know that $86,000 from your grant was sent to ATSDATA—which is a *fake* company?" Charlie said "Oh my God! I had no idea that it was a fake company." Charlie was then told that the meeting was over. He, Dave Stetler, and Mariah Moran (Stetler's assistant) immediately left the meeting room.

Once they were down the hall in Stetler's office, Charlie's cell phone unexpectedly rang.

Charlie heard the voice of his former assistant Feyi. He handed the phone to Stetler, who immediately hung up. Stetler announced that their business for the evening was over. What came of this exercise in lawsuit psychology, Charlie never learned. Nor did he learn why Feyi had called him.

This bizarre episode, along with Stetler's overall lack of action, raised questions in Charlie's mind about Stetler's role. "He acts as though he is lawyer for the other side, or he's passing on my scientific findings to Rosenbloom who is probably passing them on to Amgen," he told Amy. Stetler was opposed to Charlie's taking any actions on his own—such as telling his story to a sympathetic reporter. When we interviewed Stetler's friends, they defended his inaction, believing that he was focused on preventing a criminal indictment that he knew was a possibility based on conversations with his former colleague Joel Hammerman. Charlie found it all depressing and confusing.

In an effort to return to his research and be allowed back into his office, Charlie engaged Richard Gonzalez, a plaintiff-side employment lawyer and member of the Chicago Kent Law School clinical faculty. Gonzalez was willing to represent Bennett for a reasonable fee. He was friendly and accommodating. But Gonzalez, a newcomer to Northwestern and its policies, knew nothing of the background of Charlie's employment situation. The Bennetts were out of money. Without more research, he was not in a position to bargain effectively on Charlie's behalf. What he did achieve was Vaughn's agreement that Charlie would be granted a six-month leave of absence with pay (ironically, something he was already entitled to under university rules).

Charlie, however, would not be able to return to his research in the short term.

In early September 2009, Gonzalez received a follow-up letter from Vaughn, granting Charlie leave but attaching a series of onerous limitations:

> During your academic leave you will limit your University work to matters or projects that you can perform independently without consultation or assistance from University employees, off University premises and that will not involve any activities that are funded in whole or in part by University funds or research grants administered by the University (including subcontracts and consulting work on such grants).
>
> You agree further that you will not undertake any activities that could result in the potential expenditure of either University funds or grant funds administered by the University . . . no expense

> payments for lectures or writing except with the prior written approval of expenditures from Dr. Jonathan Licht. You will not apply for any new grants through the University. You will not enter the Northwestern University premises during the course of your leave except with the prior written permission of the Chairman or Dean. You agree that you will direct any notifications you receive or questions you have relating to your University duties to Dr. Licht.

Gonzalez told Charlie that he had no choice but to accept the conditions.

Gonzalez responded to Vaugh's powerful demands with a response letter agreeing in principle with the proposal but urging that "any inquiries regarding Dr. Bennett's status insofar as being on a leave of absence be directed to you and that and any inquiries relating to any on-going investigations be directed to counsel." When the university agreed to use Vaughn as the conduit of information to and from Bennett, the deal was set.

Vaughn's draconian conditions also made it impossible for Charlie to teach, to act as mentor, to see veterans with prostate cancer, and to do collaborative research—the activities that are at the essence of what it means to be a professor. Charlie could not continue to function professionally under these conditions. He was not even able to receive and respond to government requests for information about the progress of his own NIH, VA, and American Cancer Society grants. Throughout our investigations, we were continually surprised and confused that authorities at Northwestern University imposed these conditions without any official university finding that Charlie violated any law or scientific protocol.

Vaughn's unilaterally imposed conditions might well have provided a basis for a lawsuit alleging breach of contract and violation of Charlie's academic freedom.[5] But Gonzalez never raised the possibility of challenging Vaughn's conditions with Charlie. In addition, Gonzalez did not insist, as a condition of agreement, on Charlie's retaining his grants at whichever institution he transferred to. His answer to Vaughn was a model of timidity.

Charlie felt "overmatched." What good were his lawyers doing for him: "I wondered about lawyers in general. Maybe the whole profession was made up of suboptimally trained individuals who were not useful in a crisis."

Under the concept of academic freedom, which Northwestern University purports to accept, it is improper to penalize a tenured faculty member without an appropriate determination of guilt by an academic tribunal. It is true that the outside office of the state's US Attorney's Office was investigating Charlie for misuse of government funds in the False Claims allegation, but the specific allegations were under seal, and the state had not investigated them yet—and certainly Northwestern had not.

It is a precept of academic freedom, just as it is of constitutional justice, that a person accused is to be presumed innocent and is entitled to be informed of charges and then to respond to them. Northwestern University no doubt had procedures in place whereby a committee/administrators could investigate charges against faculty members. Vaughn's actions were a blatant violation of Charlie's academic rights. We had to wonder what form of misbehavior Dr. Vaughn feared Charlie might engage in. What harm to the university might occur if Charlie were permitted to discuss with his university colleagues the topics of his research, or if he entered his office?

The terms of the leave did not require Charlie to leave Northwestern as a faculty member, but they prevented him from communicating with his research team and his academic colleagues and even prevented him from leading the scientific programs he had developed.

Charlie feared that if he tried to return to the campus and his former office when the six-month leave ended, he would formally be charged with some other, potentially criminal acts.

Reluctantly, Charlie decided to avoid further internal struggles by leaving Northwestern.

Stetler reinforced his decision. "Good idea; start over," he told Charlie. Some of Charlie's motivations were personal: he was outraged by the suspicion and disrespect from the office staff and by the hostility from high-level administrators. Also, as his work inevitably suffered, he feared the professional consequences of staying. As he later put it,

> I was stymied everywhere. I couldn't get any work done because this administration insisted on keeping the doors to my office and lab locked. Our research group's project was really in a tight timeline. We had six months—we were done. If I didn't make it, I'd probably end up under administrative leave and I'd never get a job.

In the past, he had received indications of interest from major universities in large part because his scholarship was widely cited. He had also recently learned that he was on the short list for the deanship at the University of Illinois, Chicago School of Public Health. Those responsible for hiring at leading medical schools, Veterans Administration medical centers, and public health schools would wonder about the circumstances of his leaving and would undoubtedly check with Northwestern.

Friends encouraged Charlie to apply for a position as professor at Tulane's School of Public Health. He found the idea appealing because the Tulane School of Public Health was top tier and he had friends there and at Tulane's medical school. "But the deans of the Public Health and of the Medical School rejected the ideas based on what they heard was going on at Northwestern. They had called my division's senior persons; they just didn't want to take that risk." Charlie was fighting the battle of his academic life with his hands tied behind his back; without an official investigation by a Northwestern University committee about specific wrongdoings, he could not refute any charge another school might hear through the academic grapevine.

Charlie's former colleague Boris Pasche, now the director of the National Cancer Institute at Wake Forest University, lamented his inability to help Charlie get an appropriate position:

> I tried to recruit him both to University of Alabama Birmingham and Wake Forest just when the story broke out. So then, suddenly, I heard my colleagues saying, "Oh but did you hear *this*," "there are significant issues" and so on. It became difficult to overcome the gossip. "Oh, Dr. Pasche, there is this real problem with this guy; we cannot hire him."[6]

Charlie interviewed at the University of Miami Sylvester Cancer Center with clinicians who had trained with him at UCLA during the time of his PhD work at Yale. They were interested to hire Charlie because, for several years, he had collaborated with them on his TTP grant and his VA grants on HIV. It seemed that an offer was in the works. One day, Charlie remembers getting a letter from the director of the Sylvester Cancer Center, Fred Goodwin, who said that a follow-up telephone call with "someone at Northwestern" had not gone well. Goodwin in turn called the University of Miami counsel's office and instructed him not to hire Charlie because of "yet to be determined" legal concerns.[7] Charlie's good friend, now a dean at a Florida medical school, admitted, "Charlie, I cannot get you a job because this stuff against you is so big that my lawyers won't accept you."[8]

CHAPTER 5

Zenith: South Carolina

As the legal investigation dragged on without any formal declaration of actual criminal charges, the allegations took their toll. Depression set in. Charlie confessed: "I felt that my whole career, everything I care about, was ruined. How was I supposed to fight these rumors?"

When he was nearly out of hope, in 2010 he visited the University of South Carolina.

Joseph Dipiro, PhD, then dean of the Pharmacy Schools at the University of South Carolina and the Medical University of South Carolina, recognized that he was in a position to hire a nationally prominent scholar with a remarkable record of pointing out serious flaws in medicines that had been created and sold by pharmaceutical companies. He suggested to the State Board of Education that they offer Dr. Charlie Bennett a newly created position, one of the forty-seven prestigious SmartState professorships and also the founding director position at the South Carolina SmartState Center in Medication Safety and Efficacy. The offer was made, approved, and publicized by the South Carolina School of Pharmacy; Charlie was excited to accept. Overnight, his destroyed world was reshaped, and he could step back into the role of scientific investigator and gadfly.

The SmartState professorship and directorship positions carried with them a $6 million endowment. It allowed Charlie to continue his mentoring program and also allowed him to create a Southern version of RADAR, which he designated as SONAR (Southern Network of Adverse Reactions). At Northwestern, he had had an endowed chair at the business school and the medical school—but had to start from scratch to get the grants funded for his research.

Despite his rekindled enthusiasm, there were obvious costs in leaving Chicago's Northwestern University School of Medicine. He was separating from friends he cared for and from colleagues with whom he had shared research ideas and shown drafts. He was leaving his home and neighbors and a way of life. "Clearly, I wasn't totally happy. I lost sixteen years of my friends."

Despite the money, his office was smaller. He didn't know his colleagues, but that part was easy for him. This new position allowed Charlie to continue his mentoring program with stellar academics.[1]

When we visited the Bennetts in 2019, it seemed everyone on the laid-back, quiet campus knew Charlie. He stopped and spoke to people on the sidewalks; he approached visiting parents and assured them that, yes, he would be happy to give their entering students campus tours. His constant smile and commanding voice carried us up and into the university's president's office, where Charlie engaged in a congenial discussion of his research and goals.

The president knew all about Charlie's research—the SONAR publicity kept him informed. Down the mall at the president's gracious home, the president's wife gave us a copy of her book on the campus and Columbus' history. His new home was lovely, and his wife found a job she enjoyed; their son had entered a challenging and supportive school.

Despite all the Southern bonhomie, he shared with us that his being physically away from Chicago and Northwestern made it more difficult to respond to the legal charges that had been made against him. It was in Chicago, with its crowds and elevated subway overhead and multiple ethnic restaurants, where his academic future was in question and where it would be decided. His lawyer was there; so were his accusers; and so were his witnesses.

As a member of the Northwestern faculty, he had had a claim to a hearing at the university to clear his record and to call on members of the faculty and staff as witnesses. Now, he could not call the Northwestern University ombudsman to request a faculty investigation from Columbia, South Carolina—on the basis of events that largely took place in Chicago and on the faculty of another institution. Another problem was the spread of rumors about his reasons for leaving Northwestern and transferring to a less-prestigious institution. As one prominent physician, scientist, and former collaborator of Charlie who heard rumors about Charlie's misconduct wrote: "I obviously have no idea whether these allegations are real or true, but if they were, I would have serious misgivings about working with you."[2]

Periods of depression were inevitable as his wife Amy recounts:[3] "To me, it's a miracle that he's made it through all of this—he's threatened suicide. . . . And he's been tossed off all of the boards he was on. He was sought after, on all these committees."

She remains furious that his career has almost been destroyed.

There were no depositions of anyone and none of Charlie's defenses were ever submitted by his attorneys to the government,

> nor was the court advised of the potential conflict of interest between his attorneys and Amgen. This matter was a forced settlement by big government and institutions against someone with no means to pay for attorneys to establish the truth. There was no evidence, as my father, who was a lawyer who is now gone, said "there is no evidence."

Echoing a powerful point made by several of her husband's supporters, she kept hammering the point to us:

> The University is responsible for the administration of a grant. The scientists and the Principal Investigators are only legally responsible for the *science*. There was never an allegation that the science was bad. There were only allegations that the admins were bad—which is the University's responsibility.

From 2010 until 2014, Bennett lived and worked in Columbia, South Carolina, while his past and reputation were being evaluated in Chicago. During that time, the Department of Justice investigated him and considered bringing criminal charges against him; they also were deciding whether to join the Melissa Theis's civil False Claims case against him and the university. He had no role in any of the investigations and negotiations. Like Joseph K. of Kafka's prescient novel, *The Trial*, he was informed only that "Proceedings are under way and you'll learn everything in due course."[4]

CHAPTER 6

Details of the Government Investigation: Innocent Actions, Grounds for Suspicions

The False Claims complaint that the government unsealed in 2013 for the first time did not include scientific misconduct or personal enrichment; nevertheless, the FBI and the US Attorney's Office had launched a broad investigation of Charlie's grant expenditures.

The first person the FBI interviewed in 2009 was Melissa Theis, the administrative clerk who filed the initial charges against Charlie. When she had worked for the university, she said she noted "red flags" in the invoices that Charlie submitted. Among the details were the following:[1]

- An invoice for a Benjamin Djulbegovic appeared to contain "whited out" items. In addition, the invoice appeared to have been copied and pasted.
- Some invoices related to Charlie's NIH grants were described in a "vague and general" way.
- There were some duplicate invoices.
- Check copies were returned after deposit disclosed that the checks were endorsed by individuals and not companies.
- Some amounts requested by invoices were even numbers.
- The chief executive officer of MedRed, another company for which payment had been requested, was a former colleague of Charlie's.
- Charlie spent more money just as the grant ended.

How these details suggest that a prominent academic would knowingly file false claims is difficult for us to understand—"even numbers"? "Whited-Out" numbers?

"Duplicate invoices"? There is no explanation of how these minor details turned Theis's mind to the False Claims Act. Nor does the FBI interview report suggest that they asked her to explain the basis for her conclusion based on these "facts." (Unfamiliar with end-of-year accounting, Theis may not have realized that this is common business tactic to not lose the unspent government money.)

The FBI, like all government agencies, not only knows about, but most likely practices, end-of-year fiscal spending.

Theis had only one allegation worthy of further examination: Charlie had submitted for his grant money to reimburse money paid to ATSDATA. The FBI investigation soon realized that ATSDATA was a fictional company created by Charlie's administrative assistant Feyi as a subterfuge for embezzling $86,000 of grant money. During her FBI interview, she *admitted* that she used Charlie's grant money for her wedding and subsequent honeymoon. Feyi had sent Charlie invoices for the fictional company—during the time that Charlie had been an in-patient at Northwestern Hospital undergoing an operation for cancer in 2006. He had signed the fraudulent paperwork along with other forms she handed over to him.

In fact, Charlie had wondered about this expenditure of RADAR grant funds to an unknown company. When he was later released from Northwestern Hospital following his prostate cancer surgery, he asked Feyi to explain. Feyi assured him that ATSDATA was a research branch of the University of Illinois, which had in fact done work for him. Charlie accepted her version: "I subcontracted to this group at the University of Illinois. That's why I got confused with ATSDATA."

Feyi made out the checks to ATSDATA and, once Charlie signed them, mailed them to the address indicated on the invoice—which was in fact a post office box that Feyi had set up. It was all done smoothly. Charlie, emerging from cancer treatment and now again engrossed in his research, had no suspicion at that time that ATSDATA was a fictional enterprise because she had legitimate-looking paperwork and the administrators above her had all signed off on the payments.

Here, any rational person would have stopped all proceedings against Charlie.

If Charlie Bennett had known that ATSDATA was a nonexistent entity and nevertheless had signed off on bills made in its name, he would have been guilty of criminal activity. However, Charlie's files, which all officials read, show that Feyi deceived him. His personal ATSDATA file contains a handwritten statement by Charlie: "ATSDATA identified as U of Illinois neuro epidemiology and survey research lab & Pharmacy outcomes study group. They worked on RADAR, TTP and prostate." This handwritten note supports Charlie's contention that Feyi lied about ATSDATA. On the University of Illinois Champaign brochure, he noted that "Feyi said that ATSDATA is the billing address for Illinois Survey Research Lab."

There is no sign that any of the stolen money went to Charlie: the FBI carefully examined his personal and family bank accounts—no increases; no purchases; no investments occurring anywhere near the time of the payments to ATSDATA, or at any other time being investigated. Rather, Feyi's account showed deposits adding up to the total amount, and her wedding cost nearly the full amount.

Beyond the lack of evidence to support it, a joint Bennett–Feyi Sangoleye plot made no sense; Charlie, who had huge grants and a large income, would not endanger his career and even his freedom for such a relatively small amount to permit someone, with whom he had only a limited professional relationship, to steal money for her honeymoon.

When the FBI interviewed Feyi's friend and former colleague Quishun Elrod, they learned that she had visited Feyi at home, believing that Feyi's husband had been aware of "what was going on"; Feyi told her that she had been "suspended" because of ATSDATA withdrawals. Feyi confessed to her friend that the ATSDATA fraud "was me, but in my husband's name."[2]

It's not surprising to those who know Charlie that he accepted Feyi's description of ATSDATA. Charlie's focus on science is intense and overwhelming. As Otis Brawley, then the chief medical officer for the American Cancer Society, told me, "Charlie is meticulous in science, but he may come to a meeting with his fly open."[3] A similar comment was made by Professor Elizabeth Calhoun: "Charlie is a genius, but his shoes may not match."[4] Steve Rosen noted Charlie's occasional mistakes of detail and, like Otis Brawley, attributed it to Charlie's focus on achievement:

> I think Charlie would admit a fault of his, is that he's sloppy. Sloppy in the sense that he's a big-ideas person, so he doesn't focus on what many would consider are dotting the *i's*, crossing the *t's*. He's much more concerned with what can be accomplished. He trips around a bit in his own world. A result of that is that the buttons of his shirt are not aligned.[5]

Even though the US Attorney's Criminal Division eventually decided in 2013 that it did not have enough evidence for a criminal indictment of Charlie but did of Feyi, some people in the office continued to have strong suspicions. Of all the expenditures that Theis listed as suspicious, the ATSDATA one was surely the basis for the "serious thought" that Assistant US Attorney (AUSA) Joel Hammerman gave to bring criminal charges against Charlie.

Feyi's initial FBI testimony holds the only suggestion of Charlie's possible involvement in any fraud. According to the FBI's summary of that first interview in May of 2010, Feyi said Charlie had always known about the theft and even would hide it for her:

> With respect to ATSDATA, Bennett **knew that the company was false since 2007** when he and Sangoleye discussed it. Bennett indicated that grant money would need to be spent. Sangoleye would have to submit invoices, but Bennett would "take care of her." For the first ATSDATA invoice, Bennett and Sangoleye agreed that, if anyone checked or verified this company, they would indicate that ATSDATA was related to the University of Illinois at Chicago (UIC). They had decided on using UIC because Bennett had many relationships with them, so the mention of the name would seem believable. [emphasis added]

Charlie's relationships with the University of Illinois Chicago School of Public Health (UIL) made them a good choice for Feyi's explanation of the expenditures to Charlie. It did not make a good excuse for Charlie, if anyone checked, because officials at that institution could quickly establish that it did not exist.

Noteworthy in her statement is her admission that Charlie is not a principal in the scheme to defraud the government. Nor does he share in the benefits. Even in her initial version to the FBI, Charlie has "agreed" to protect her only if she is under suspicion.

But then, interviewed a second time three days later, Feyi told a more confusing and less incriminating story, that she "believes" Charlie knew about it:

> It was Sangoleye's idea to create ATSDATA to receive grant money. She does not recall creating any more than 10 ATSDATA invoices. The total amount received from these invoices was $90,000. Sangoleye does not have any of this money left. She used it all to plan and pay for her wedding, Sangoleye **believes** that Bennett knew late last year that ATSDATA was fictitious. [emphasis added]
>
> At one point, when Bennett inquired about invoices, he asked about ATSDATA. Sangoleye informed him that ATSDATA was the company that they agreed to describe as related to the University of Illinois at Chicago (UIC). Bennett had no reaction to this statement. Sangoleye does not think that there are any other hints that Bennett knew that ATSDATA was fictitious, other than the mention of UIC.

Feyi's second version is far less certain than her first. She "believes that Charlie knew late last year." He had no reaction when she reminded him of its fraudulent nature, and she doubted "that there were any other hints that Charlie knew that ATSDATA was fictitious." Also, according to Feyi, Charlie seemed to have forgotten about his participation in the illegal scheme, which put his finances, his liberty, and his career in jeopardy, until Feyi reminded him of it.

It is difficult to imagine a legal conviction based on Feyi's self-serving testimony concerning Charlie's knowledge of ATSDATA.

Feyi did suggest other misbehavior by Charlie in her interviews. She cast doubt on the legitimacy of payments to Professor Elizabeth Calhoun. In her interview, "Sangoleye advised that 'Bennett and Calhoun were very close . . . beyond colleagues.' Sangoleye suspected no work was actually being performed by Calhoun."

Feyi had no evidence to back up her suggestion of an affair or her suggestion that Calhoun did no work. Indeed, Calhoun is an odd person to accuse of not working. Now a professor at the University of Arizona (recently appointed as a vice president), her faculty profile points out that "Dr. Calhoun is author of more than eighty peer-reviewed articles in scientific journals covering the impact of health and environmental disparities to disadvantaged neighborhoods. She is also author or coauthor of several book chapters."

It is easy to understand why Feyi was willing to accuse Charlie of impropriety. The practice of government district attorneys in general, and US attorneys in particular, is based on plea bargaining to convince lower-ranked suspects to implicate higher-ups by offering them sentence leniency in return.

Lindland nor the FBI ever directly questioned Charlie about Feyi and ATSDATA. We are quite certain that if experienced labor arbitrators had heard her testimony, they would have promptly discredited Feyi's testimony about Charlie's involvement in a scheme purely for her benefit. No experienced fact finder would have accepted her tale. She discredited herself with her contradictions, her admissions, and her bank account.

Dr. Roy Poses, president of the Foundation for Integrity and Responsibility in Medicine (FIRM) and clinical associate professor of medicine of the Alpert Medical School at Brown University, found Charlie's punishment bizarre: In his *Health Care Renewal* blog,[6] he exposed Northwestern's role under the subheading, "Dr Bennett Found Irregularities, Sangoleye Pleaded Guilty, Nobody Noticed":

> It's totally disproportionate; if he actually submitted excess travel and stuff like that, he just should have been slapped by Northwestern. And if it somehow got by Northwestern officials and staff and they authorized it, the government should have slapped Northwestern . . . as much too loose, and maybe written

> him a letter saying, "No, no, no." And maybe, at most, asking him to pay it back.[7]

In 2013, the Criminal Division of the US Attorney's Office finally realized that they had no evidence to support their conclusion that Charlie knew of the fraud or profited from it. Let's follow what happened to Feyi. She did not spend a day in jail; however, on June 25, 2013, she was criminally charged with Violation of Title 18: U.S. Code, Section 641. (see Appendix A). In March 2014, the United States District Court of the Northern District of Illinois issued a Judgment in a Criminal Case against Feyifunme Sangoleye responding to her guilty plea. Feyi pled guilty to creating ten false invoices for a fictitious corporate entity (ATSDATA), which had caused the NIH to pay $86,000 in 2007 and 2008 to her fraudulent entity.

What happened to the guilty Feyi? She was required to pay back, over a lengthy time, the $86,000 that she had misappropriated from the government and university (10 percent of her net monthly income). She also had to commit to six months community service, participate in a mental health treatment program, and not incur new credit card debt.[8] No publicity, no public shaming. She has risen from those ashes, however. According to her Linkedin profile, 2002, she is working as a nurse at the University of Chicago.[9]

Nevertheless, in a 2018 phone call with us, Kurt Lindland repeated his claim and still insisted that, with respect to the ATSDATA fraud, Charlie was "in on it"—even though he never questioned Charlie.[10]

In addition to the ATSDATA focus, the FBI was also interested in the potential criminal issues raised by Theis of payment to Charlie's cousin Earl E. Ripling for web hosting and design. Charlie had hired his brother Paul and his cousin Earl E. Ripling to establish a website for RADAR. Charlie's desire to set up a website was understandable. A website made the steady flow of information developed by RADAR easily accessible to physicians and scientists, and the website made it easier for physicians to report on their patients' adverse drug experiences. But finding the right person or entity to establish a continually changing scientifically accurate website turned out to be far more difficult than Charlie anticipated.

After several failures, Charlie realized that he might have the needed expertise in his own family. His brother Paul had been a website developer since 1990, and most recently had set up a professional website in collaboration with a German corporation. Charlie and Paul agreed to do the job—according to guidelines that Charlie established. Paul Bennett and his cousin Earl Ripling quickly established an active website. Charlie brought them to the university to meet Steve Rosen and members of the staff, and he notified university personnel of his decision. He had in fact sent their names to the provost for approval.

However, years later as he reviewed for us the litany of charges he fought, in hindsight he regretted entering into an arrangement that gave the appearance of nepotism to grant administrators and investigators. Even though he had introduced the two to university officials, he later understood about the appearance of impropriety and that he had unwittingly violated some anti-nepotism rule of the government's.

Throughout 2009 to 2013, the FBI persistently continued to interview staff and others who worked on Charlie's research projects. Among those was Alice Camacho. Much of the interview involved Camacho's suspicion of Charlie's consultants and subcontractors. The FBI recorded her suspicions about both personnel and expenditures:

> *A colleague who recorded V.A. patients*—There were rumors that she and Bennett had a personal relationship. Her research seemed different than RADAR. It involved research into veterans' history. [Charlie worked for the VA.]
> *Ben Djulbegovic (Djulbegovic)*—These invoice amounts seemed very high.

> Camacho wondered why a subcontract was not set up since they seemed to be using another institution's research. Generally, consultants work at home; they do not necessarily get their names published. Their invoices would typically be around $10,000. Camacho does not know what work Djubegovic did.
>
> *MedRed*—Camacho did see a justification memo that was given to Licht by Bennett. It did not seem sufficient to Camacho.
>
> *Calhoun*—Calhoun is a subcontractor who started last year. Camacho had heard rumors of a personal relationship between Bennett and Calhoun. Camacho had informed Bennett that he would need NIH approval for additional coordinators.

Also interviewed in 2009 was Northwestern University administrative assistant Tricia Larson. According to the FBI transcripts of her interview, she found Charlie's expenses, like his meals, "excessive" as compared to other Northwestern scientists. She mistakenly claimed that he bundled vacations and professional meetings.

The FBI investigation regularly involved nonscientists, like Camacho and Larson, evaluating the appropriateness of the use of grant money by credentialed scientists.

Of course, the allegation that Charlie had requested reimbursement for vacations, if true, would have been the basis for either a False Claims Act or a criminal indictment. But no one ever produced factual evidence to support the charge. The origin of this accusation remains hidden in the web of office gossip that for a time engulfed the Hematology and Oncology Division of Northwestern. A detailed investigation of Charlie's requests for payment for work-related trips conducted sometime later (by US Attorney's Northern District of Illinois investigators) failed to reveal *any* requests for vacation-related payments.

It is interesting to contrast the treatment of Charlie with that of Dr. Bala Hota, an Illinois hospital physician who "improperly spent" $248,322. Investigators uncovered the doctor had used grant money for "personal benefit." He resigned his position in 2014 and repaid the money in 2017. He was not charged with a crime nor disciplined by the Illinois Department of Financial and Professional Regulation.[11] In contrast, somehow the Illinois Department of Financial and Professional Division of Professional Regulation required the Bennett family to pay a $35,000 fine to keep them from suspending his medical license pending proof of innocence—with no admission of liability on Charlie's part. Their agency's 2013 actual filing was not opened for review by Charlie and his attorneys until 2016. Feyi's guilty plea was on record. Charlie had not then, nor had ever, been found guilty of fraud. The timing of the payment and uncovering of the 2013 complaint is also suspect—Charlie didn't see it or sign it until 2016.

Government agents interrogated many of Charlie's friends and coworkers concerning small expenses that Charlie had sought repayment for through his RADAR grants.[12] Most had only a hazy memory of the meetings. Typical of many in its focus on minor expenses was the following HHS interview summary of Denis Cournoyer MD, professor of hematology, at the McGill University School of Medicine: "Cournoyer is familiar with Charles Bennett. They worked on scientific matters together. . . . He has done some work related to specific aspects of RADAR. However, Cournoyer does not recall whether he was compensated for this work or for work related to TIP."[13]

One of the most accusatory interviews was with Charlie's division chief, Jonathon Licht.

The summary of the interview suggests that Licht developed a deep suspicion about Charlie:

> [Licht] would have considered Bennett's lies and actions to be fraud (e.g., the fact that Earl Ripling was Bennett's cousin). At the time, Licht presumed Bennett's innocence and did not suspect anything. However, Bennett's answers were not sufficient.

> . . . At this point though, Licht had still not talked to Bennett in person.[14]

The quoted portion, like the entire interview, is accusatory without basis. Licht assumes that the hiring of Earl Ripling, being Charlie's cousin, is evidence of fraud but does not tell the FBI which lies he was referring to. After the investigation began, the division chief never did talk with Bennett in person.

That's it. A broad review of the notes from the FBI interviews reveals that none of these interviews produced convincing evidence of either illegal or inappropriate behavior on Charlie's part. The government investigation into Charlie's grant management at Northwestern dragged on for many years after Charlie left. AUSA Kurt Lindland, still in charge of the investigation, never changed his conclusion that Charlie was in on the ATSDATA scheme.[15]

CHAPTER 7

Dr. Bennett Finds a New Cause: "Flox"

In March 2010, Charlie, lacking professional purpose, was depressed. He was not allowed to meet with collaborators; was prohibited from seeing his VA cancer patients; was kept from national committees and speeches. He frequently fielded calls from his former colleagues and friends, saying in effect: "Charlie! The FBI is at my door . . . what do I do?" He always answered the same: "Open the door. Tell the truth."

Hoping to lighten his mood, his South Carolina research program director suggested that he attend an FDA meeting in Washington about partnerships between the FDA and local communities.[1] Perhaps Charlie might be given a role in one of the programs that would likely develop at the conference. On the lookout for a new project, Charlie agreed to attend.

The meeting began as most government/academic meetings do, with an exchange of earnest clichés stressing the importance of government/private sector collaboration to minimize risk and improve safety. How impersonal and meaningless. Charlie found it difficult to listen. Nothing new.

But Charlie's interest was suddenly aroused when, during a subsequent question-and-answer period, John Fratti (a former pharmaceutical drug representative) described how the widely prescribed J&J antibiotic Levaquin had severely damaged his health.

> I have been disabled for over five years from Levaquin. I have documented nerve, tendon, and central nervous system damage. I have had many diagnostic tests on my brain and body that show the damage from Levaquin. Prior to my Levaquin injuries, I was, ironically, a pharmaceutical sales rep. . . . I am currently on many different prescription pain medications for my injuries.[2]

Fratti's list of injuries was lengthy and serious:

> I hadn't been able to see right for over five years. The damage to my vision is but one of many central nervous system injuries. I also have . . . chemical sensitivities, tremor in my head, severe insomnia to the point that I have to take a drug in order to sleep. I am unable to concentrate for extended periods of time. I have many other symptoms from Levaquin.

Yet another major drug advance by J&J, yet another serious and hidden adverse reaction. The problem dated back to the mid-1980s when two new blockbuster drugs were developed to treat serious infections from Levaquin (J&J's 1996 FDA approved drug) and Cipro (Germany's Bayer 1989 approved drug). Use of both drugs for sinusitis, pneumonia, urinary tract infections, anthrax, and the plague spread rapidly worldwide. Wonder drugs, their positive effects obscured adverse reactions for a long time.[3]

After the formal meeting, Charlie charged forward and introduced himself to Fratti: "Your story was one of the most moving and powerful that I have ever heard. I have studied adverse drug reactions for many years, and I would be happy to work with you on the problem of fluoroquinolones, such as Levaquin."

Fratti responded enthusiastically. He told Charlie that he was part of a group with three other fluoroquinolone sufferers who regularly conference-called each other to discuss their experiences and those of other victims. They regularly referred to someone who experienced adverse reactions after fluoroquinolones as having been "Floxed."

Through Fratti, Charlie met the other three people in the group: Linda Martin, a business executive; David Melvin, a former state police officer; and Alan Redd, an anthropologist and assistant professor at the University of Kansas. The foundation of the group was a website, myquinstory.info, that David Melvin developed to vent his frustration at the incompetence and lack of concern by the medical profession.[4]

> I developed a case of epididymitis and some pain and swelling, and I went to my doctor. My doctor was, kind of, concerned and sent me to a urologist, and the urologists gave me some Levaquin. I had a fairly large dose of Levaquin.
>
> I went ahead and finished the course. And then, my problems just slowly started, beginning with severe muscle pain. This led to numerous visits to the urologist and rheumatologists. Everything kept coming back negative; they would do EMG after EMG. And my symptoms progressed.

> So, I went to the University of Chicago. The professor that I talked to up there, who was in charge of the neuropathy clinic, looked at me and said, "We got, really, nothing for you. We're going to give you a prescription of Lyrica to treat your pain. If it gets worse, come back and see us." She looked right at me—by this time, I had done a little bit of research. And I said, "It looks like it could cause neuropathy." And she looked right at me, and said, "*not* by any papers I've read."
>
> I went back and, as a catharsis, I went online and put up a blog, initially, just to vent my frustration.

After his website went up, Melvin began to hear from other fluoroquinolone victims, and the idea of forming a group occurred to him. In our series of phone interviews, he told us that he realized that many of the victims had interesting educational backgrounds and, if they investigated together, "we might achieve something."

The group had no name, no formal structure, and no clear purpose. Each of the members sought to learn more about what happened to them, and all desired a measure of justice for themselves and other "Floxed" victims. Collectively, they realized that a lawsuit would be too costly and time-consuming and the result too uncertain.[5]

In an unusual move, John Fratti had bought stock in J&J so that he could attend and speak regularly at shareholder meetings, "very simple, very respectful."

Shortly after meeting Fratti, Charlie set up a five-way phone call to talk with the four activists. They presented their individual stories of constant pain and major life changes; it occurred to him that the side effects of fluoroquinolones were a major pharmaceutical problem that had not received adequate critical focus since ciprofloxacin entered the United States' market in 1989.

Charlie was stirred to action, investigating a major class of drugs—again. He would be once more challenging accepted medical wisdom.

If it turned out that the Floxed group's charges were accurate, he would be exposing devastating side effects. This had echoes of his successful work on Ticlid and Plavix, and also the EPO battle that caused his Northwestern University downfall. Dangerous perhaps, but also important.

Charlie learned there was scant formal investigation into fluoroquinolones—but many online personal testimonials. The medical profession largely ignores these for understandable reasons: patient testimonials are rarely trusted or treated as significant evidence by drug companies, scientists, and the FDA—all of whom are legitimately concerned that individual details may be exaggerated, misinterpreted, used as a basis for litigation against physicians or drug companies, or falsified.

Charlie immediately turned to his University of South Carolina SONAR group that included almost all of the old RADAR group (with the notable exception of Dennis West). He presented the two-sided issue: how can the established benefits of Levaquin and Cipro be preserved and yet the side effects minimized? For instance, he theorized, perhaps markers could be discovered or developed for those patients who might be at risk for serious toxicity following Cipro or Levaquin. Charlie invited Linda Martin, from the Floxed community, to join in SONAR's weekly telephone meeting. Charlie's and SONAR's involvement shifted the focus of discussion within the small Floxed group to developing an FDA presentation.

In the meantime, more and more individual stories about lives ruined appeared online. Most of the patient testimonials were similar to John Fratti's story. Some had more detail, and many reflected more anguish. Typical is Laura's story, posted on fluoroquinolonestories.com. Her problems started when she went to an after-hours health clinic:

> I had a urinary tract infection and was prescribed Cipro. Seven days into the ten-day script, I felt a funny tingling sensation in my right leg. That progressed to both legs and both arms and

> went from tingling to a very painful cold, burning. I am now 49 years old and the cold burning is STILL there. In addition, I've got severe tinnitus and hyperacusis. Cipro has completely devastated my life. I have skin-biopsy-confirmed small-fiber neuropathy and am severely intolerant to cold temperatures and loud places. Some of the other issues I've had have been a separated shoulder, frozen shoulder, hip pain, detached retina, and anxiety. I was not warned about ANY side effects by the doctor or the pharmacist and was told the most important thing was that I finish all the medication. The initial tingling was so odd I never associated it to the Cipro. It took a few weeks for it to become a painful burning and to spread to all my limbs and also my face at times.[6]

Stories like this elicited many supportive comments and many suggested therapeutic suggestions from fellow fluoroquinolones sufferers. Laura concluded with a widely shared belief that fluoroquinolones should be prescribed only for serious, potentially fatal infections.

The website continues to post personal stories of the devastating effects of these drugs.[7] Hearing and reading about these life-changing consequences of fluoroquinolones, Charlie began studying them intensely, and he encouraged younger colleagues to work with him. The person most influenced by his urging was Dr. Raja Fayed, a physician from Syria and an assistant professor at the University of South Carolina who had worked at the University of Illinois Chicago when Charlie had been at Northwestern. Both now in South Carolina, Raja and Charlie met regularly to discuss ways in which Fayed might undertake research and experiments related to drug safety. They decided to collaborate on a laboratory study of the impact of fluoroquinolones on mice.[8]

A few months after their initial discussion, Martin remained worried that there had been no public statement issued by SONAR about the harmful effect of fluoroquinolones:

> I called Dr. Bennett before Christmas, and I begged him, literally, "We've got to do something. You know, we've got all these people who are desperate for some information." And Dr. Bennett said, "Okay . . . I'll get back to you." Three or four months later in the spring, I got a draft of a research paper that Dr. Bennett had done with Raja. They actually had developed a lab research study with rats using Cipro. It was such a shock to me because I had no idea they were actually moving forward, and it was the nicest thing anybody had done for this huge community of damaged people.[9]

Inspired by Charlie's actions, Linda Martin became increasingly involved in the scientific investigation of dangerous drugs, including becoming an active participant in SONAR discussions.

> I participate in the SONAR calls every week. The science of it is often way over my head, but I learn a lot—there's probably between ten and twenty people participating, depending on the day. But the discussions are incredible with things they are discovering and the approaches that they're using to identify issues before they end up like the massive numbers of people who have been damaged by Levaquin.
>
> SONAR is catching problems earlier than others have in the past. Charlie really is doing a remarkable job that the FDA should be doing,[10] that the drug companies should be doing, but Charlie is the one doing the research.[11]

Martin also discovered that battling Big Pharma carried with it many different types of risk.

She told us of a conversation between Charlie Bennett and Dr. Beatrice Golomb, a fluoroquinolone researcher from the University of California San Diego, who is also a critic of drug company misfeasance.[12] During that conversation, they referred casually to threats from drug companies that had reportedly been received by academic clinicians:

> I got them together, Dr. Golomb and Dr. Bennett. This was early on, and we were sitting there waiting for another doctor to join us when Dr. Golomb and Dr. Bennett started talking about who had gotten threatened lately. "Well, so-and-so got a death threat, and so-and-so got a death threat," and this was so new to me. I sat there just with my mouth open, because it was just like they were saying, "Pass the donuts" or something. It was so casual. And I thought, "What are you talking about?" This was early on before I had any clue. These drug companies are vicious, they're ruthless, way beyond anything I ever could have imagined as a private citizen before I got involved with this.[13]

Charlie, and in a 2020 telephone interview, Dr. Beatrice Golomb both confirmed the accuracy of Martin's story. Dr. Golomb added that such threats are common in the world of pharmaceutical investigations.[14]

The FDA, the Department of Health and Human Services, Public Health Service, Food and Drug Administration, and Center for Drug Evaluation and Research, authored "Disabling Peripheral Neuropathy Associated with Systemic Fluoroquinolone Exposure" in 2013.[15] The FDA discussed not only fluoroquinolone-associated neuropathy, but also post-fluoroquinolone mitochondrial toxicity, which is associated with ALS, Alzheimer's, and Parkinson's diseases.

However, neither Charlie nor the Floxed community was satisfied. They now focused their activity on getting the FDA to change the warning labels to include a Black Box warning[16] about the *potential for neuropsychiatric toxicities* following use of Cipro or Levaquin. Working together with Linda Martin, Charlie developed and sent two citizens' petitions to the FDA in June and September of 2014. The first petition requested that the FDA require J&J and Bayer to add to the warnings of "potential mitochondrial damage."[17]

The second petition, based on the FDA's own fifteen-year datasets, requested that the FDA require a black-box warning patients about fluoroquinolone psychiatric side effects, including suicide.[18] [See final chapter of book and Charlie's continuing fight.] The FDA rejected the first citizen petition on mitochondrial toxicity as a possible mechanistic pathway outright, and it announced that it needed further time to study the second citizen petition related to the clinical manifestations of severe psychiatric toxicity following Levaquin use.[19]

Charlie announced the results of his fluoroquinolone research in 2015; when interviewed, his public statement pragmatically stressed that SONAR was *not* seeking to eliminate the use of fluoroquinolones:

> Levaquin and Cipro are important drugs . . . especially to treat secondary infections in patients undergoing chemotherapy. Every drug has side effects, . . . If the drug has a good risk-benefit profile, then it should be on the market. . . . Most doctors don't know these side effects exist. On top of that, the most common use of the drugs is off-label. The drugs are only appropriate when used as directed.
>
> Levaquin, Ciprol—these are billion-dollar drugs, but the FDA has the authority and responsibility to ensure they're labeled correctly.
>
> In my life I always wanted to be in a position to make a difference. I think that's what we're doing. We have saved thousands of lives and billions of dollars. That's a good day's work.[20]

Following the two petitions and additional research, the 2015 FDA's Anti-microbial Drug Advisory Committee held a one-day meeting. Of particular importance was that the FDA defined a new diagnosis: Fluoroquinolone-Associated Disability (FQAD). After hearing from all the stakeholders,[21] the twenty-one-member FDA Advisory Committee *recommended* that relevant pharmaceutical manufacturers of fluoroquinolone antibiotics improve safety descriptions on product labels.

After four years, the FDA approved in part the second citizen petition dealing with severe neurologic and psychiatric toxicities.[22] Janet Woodcock MD, then director of FDA's Center for Drug Evaluation and Research, informed Charlie that manufacturers of all six FDA-approved fluoroquinolones would be *required* to identify six of twelve SONAR-identified psychiatric syndromes (agitation, delirium, disturbances in attention, memory impairment, disorientation, and nervousness) in product labels. And product labels would now include a new subsection, "Central Nervous System Disturbances," describing these diagnoses.[23]

These changes were certainly a positive step, but not enough for the persistent Flox group. As stated by Martin: "They didn't put cardiac issues in the black-box warning; they didn't put neuropsychiatric in the black-box warning; and they didn't use the word 'disability.' They watered it down."

Most people with toxicity do not read scientific journals. Charlie and Martin realized that, to save lives, they needed to inform the public about their research information so that people could decide on their own about the benefits and drawbacks of these drugs. Plus, the FDA needed to be prompted to act.

Charlie and Martin developed a nationwide campaign to encourage quicker and more decisive FDA action. The campaign centered on local television interviews with Charlie, plus local people who believed they had been seriously injured by Cipro or Levaquin. Martin coordinated the patients to be interviewed on local news shows and gave the news anchors SONAR's findings and other patient stories. Martin proved be efficient at public relations, in part because of her own story, in part because she constantly read other sufferers' stories, and in part because she joined the SONAR mission to provide patients with the best scientific information available:

> It started off by me reaching out to reporters; but, once we had a couple of stories going, then Charlie's name was out there, and people started contacting him directly. Charlie would then notify me and say, "This reporter called me from Detroit and wants to do a fluoroquinolone story." I would ask Terry [a patient who had recently joined the Flox working group], "Can you find somebody in Detroit that would be willing to be interviewed?" Terry would give me the name of a patient who had been Floxed, and I would give background to the reporter and arrange for a time for the reporter to interview Charlie, and a separate time for the reporter to interview the patient. That's how the whole model worked—we always needed to have a local patient. That was the reporter's requirement.[24]

Charlie almost never turned down an interview, traveling from city to city on funds that had been allocated to Charlie's SmartState Center for Medication Safety and Efficacy and the related SONAR initiative program at the University of South Carolina. The safety program allowed him to deliver warnings about the possible dangers of fluoroquinolones and the need for action to disseminate this information to physicians.

Between 2015 and 2017, Charlie did more than two hundred interviews at television stations in virtually every major city in the country; he conducted roughly half of the interviews in person at the local television station's offices. Charlie went to one interview immediately after his mother's funeral in Pittsburgh in February 2015; another took place a month later at the funeral home where his father-in-law was being buried in Milwaukee.

At the end of many interviews, the station would ask a spokesman for J&J or Bayer to respond. According to Martin, the representative would usually offer a generic response, sending a note stating that the company had complied with risk-benefit requirements, that the FDA had approved of the products' use, and that no change in the product label was needed. Nevertheless, the news articles became a rallying point for patients suffering with adverse reactions to fluoroquinolones. For instance, Atlanta's consumer reporter Jim Strickland's Emmy-winning Channel 2 Action News (WSB-TV) show featured Charlie and the active Flox group.[25]

The following story is similar to many others that anchors picked up and reported, focusing on patients' experiences:[26]

> DALLAS (CBS 11 NEWS)—When you go to the doctor, you want them to give you something to make you feel better. But for some patients, the cure may be much worse than what they came in with. The pill called Levaquin is part of a group of antibiotics known as "quinolones." While these already carry a serious warning about side effects, some say they could also be responsible for death . . . says researcher Dr. Charles Bennett. "We're talking about going to a physician's office, having a little bit of a sniffle, coming out with an antibiotic and then shortly thereafter having these kinds of problems."
>
> Problems like a ruptured tendon or nerve damage. Martin Landmon says it happened to her. Says Landmon, "As I was walking through my house . . . I was thinking, there's something wrong. I mean my foot hurts, my ankle hurts, my shoulder started hurting." . . .
>
> Landmon and the others received a powerful antibiotic called Levaquin. Landmon had it while in the hospital for kidney stones. She says she didn't receive any kind of warning about potential side effects.
>
> Levaquin does carry what's called a black-box warning. It tells patients of a specific side effect—ruptured tendons. The Food and Drug Administration says it can also cause nerve damage. But Dr. Bennett claims there could be bigger problems: "What we're talking about is much more serious. . . . 30,000 deaths potentially."

Bennett says when a doctor writes that prescription for Levaquin, they should make a point to ask whether they really need it. Landmon wishes she had had that chance. "It's like my whole body is crumbling from the inside."

The main purpose of Martin and Charlie's media campaign was to put pressure on the FDA *to require* drug companies to update their product labels, but the interviews and articles also had another, less easily anticipated achievement: they helped raise morale among the "Flox" patients. Linda Martin told us, "Charlie has no idea of how many people think of him as their hero. He was their voice. Floxed patients often feel lonely and abandoned. Charlie let them know that they had important support."[27]

Various internet groups made sure that Charlie's interviews were widely seen by posting them on social media. Those who knew of impending television interviews often gathered to listen and discuss follow-up actions. They recorded and replayed interviews for those who missed the initial broadcast. The sense of optimism and solidarity by patients and their families grew.

In 2016, the FDA announced that it would henceforth *required* that J&J and Bayer add updated boxed warnings and disseminate medication guides that advised patients and providers about serious side effects with fluoroquinolones. Additionally, they stated that the risks generally outweigh the benefits for patients with uncomplicated cases of sinusitis, bronchitis, and/or urinary tract infections when other treatment options were available.

The FDA *recommended* that J&J and Bayer advise physicians to prescribe fluoroquinolones to patients with these conditions *only* when fluoroquinolones were a last resort. Then, after a four-year study, the FDA approved the second citizen petition in part, dealing with severe neurologic and psychiatric toxicities.[28] Charlie and his Floxed group were elated, with Charlie calling this FDA announcement a "huge paradigm shift."

But, again, he called for further action: "What needs to happen is the patient, doctor, and pharmacy all need to sign off that they've been educated about the drug beforehand." When he and Linda described these attempts to get our own doctors to sign off on labeling, our imaginations couldn't breach the gap between the label and reality. When we recently broached the subject with the possibility of an educated sign-off, he deflected the question and said that "they" never tell doctors anything.

Charlie said the FDA should also consider updating fluoroquinolone warning labels to consistently address the risk of suicide: "The changes in the package insert do not consistently address suicide as a potential toxicity with all FQS. We've identified 122 patients who have committed suicide following initiation of Cipro or Levaquin—and 45 percent of them DIED within two weeks of starting the drug."

Charlie's efforts also affected the status of Floxed individuals in Europe. In 2016, the German journal *Kontraste* published a scientific article on the dangers of fluoroquinolones focusing particularly on Cipro produced by the German company, Bayer.[29] It pointed to the new US labeling rules as a model, and it referred to Dr. Charles Bennett. The German article was followed by action including an extensive public hearing. In September 2018, the European Medicines Agency (the equivalent of the FDA in Europe), advised the medical programs of the then twenty-eight European Union nations that "fluoroquinolone use be permitted *only* for persons for whom no other antibiotic is possible."[30]

Despite these successes, the unexpected consequences of and adverse reactions to fluoroquinolones continue to this day.[31]

The addition of patient input and their medical concerns has made SONAR into a more effective, multifunctional entity that encourages the patient involvement in recruitment for genomic studies of drug toxicity—an upgrade of the work that Charlie did with RADAR at Northwestern University between 1998 and 2010.

This resurgent episode shows Charlie at his best: listening to patients to discover an important danger caused by popular medications; filing petitions to activate the FDA; giving talks and being interviewed at major cities across the country. Those who hoped Charlie would be silenced or rendered impotent by leaving Northwestern University turned out to be widely off the mark.

In any reasonable world, this should be the end of this book. But remember—Assistant US Attorney Lindland was determined to prove that Charlie had committed fraud, and the legal case was relentlessly moving forward.

CHAPTER 8

Northwestern Settles Whistleblower Suit; Charlie Stands Alone

From 2010 to 2013, while Charlie was concentrating on Flox, the US Attorney's Office continued to focus on Charlie's possible civil or criminal violations of the False Claims Act.

In March of 2013, Stetler called Charlie, informing him that the criminal investigation by the US attorney's Office against Charlie was ended. "Go and live your life," he said.

Before Charlie had time to celebrate, Stetler called again to tell Charlie unfortunately and unexpectedly that his legal problems had not, in fact, been solved. Kurt Lindland and the relator intended to pursue civil penalties against him under the Civil False Claims Act. Lindland simply could not accept the conclusion that Charlie was innocent of criminal behavior.[1]

More bad news followed. Later that year, Northwestern and the government reached a deal in the unsettled False Claims Act case. The University agreed to pay nearly $3 million in return for dropping the official oversight and negligence charges against it and Dr. Rosen. To the dismay of Charlie and his family, his friends, and his professional colleagues, the settlement did not include Charlie despite the weakness of the case against him and despite the benefits that flowed from the very grants about which he was questioned. "This mystery case was a snowball in action," observed Steve Rosen. "Once the government demanded answers, and posted the penalty, Northwestern just rolled over and rolled over on Charlie to cover itself."[2]

According to James McGurk, former assistant US attorney (AUSA), who represented Charlie during this period, "Everything should have been closed together." McGurk pointed out that if the settlement had included Charlie Bennett, it would have had significant advantages for all sides in this complex dispute. For the relator, it would include payment and closure; for the government, the ability to shift energies toward more significant concerns; for the university, it would mean closure and protection against continuing legal vulnerability; and for Charlie, it would mean getting his life back and continuing to focus on issues stemming from adverse drug reactions.

The failure to include Charlie in civil settlement was noted by *The Cancer Letter*,[3] which was openly suspicious and indirectly suggested that Charlie's exposure of EPO was to blame: "Making enemies is an exhaustively studied side effect of probing the safety of cancer drugs, and during much of the period in question, Bennett was producing more than his standard quota of foes." As we chronicle later, several of the Northwestern board of directors had a direct connection to the largesse of Big Pharma.

Charlie was dismayed by the result and angry at being cut adrift by the university; he now faced the prospect of civil litigation with the recently unsealed qui tam case. He turned the focus of his anger toward David Rosenbloom—the lawyer who negotiated the settlement on behalf of the university, and toward David Stetler—whom the university had paid many thousands of dollars to represent him. Charlie has long insisted that both David Stetler and David Rosenbloom had a conflict of interest that made their representation of him inadequate; Charlie revealed that outrage and suspicion in a letter to his subsequent attorney Rob Hennig:

> I only discovered that Rosenbloom was working on the Amgen case when I reviewed your settlement with Amgen [Peter Winn's case] in 2013 and identified Rosenbloom's signature at the end of the document. Rosenbloom had met with me in person several times, and was well aware of my work with Amgen.

While Charlie's attorneys remained essentially inactive, Lindland continued his efforts to build a False Claims lawsuit against Charlie. One major facet of his efforts was hiring the accounting firm of Gould & Pakter (G&P)[4] to evaluate Charlie's claims for grant reimbursement. G&P issued a concluding report, finding roughly $500,000 that Charlie had claimed for reimbursement under his federal grants were "inappropriate for payment based on standard accounting procedures."

When we examined the report line by line with Charlie, we learned that the report labeled some of Charlie's most valuable research as "inappropriate expenditures."

G&P applied accounting rules as though failure to follow them constituted a False Claims violation. That is incorrect. For instance, G&P included in their list of misspent consulting money $31,250.00 paid to Dr. Benjamin Djulbecovic, perhaps the best-spent and least controversial use of Charlie's grant money. As pointed out in the previous chapter,

Ben was responsible for the key conclusion that EPO created mortality risks. Their collaboration was the basis for Steve Rosen's conclusion: "Charlie Bennett saved more lives than anyone in American medicine by his major role in demonstrating the dangers created by the billions of dollars of erythropoietin medications, sold by the pharma giants Amgen and Ortho." G&P accountants found this consulting expenditure improper—perhaps because Charlie's hiring of Ben did not meet their criterion of arm's-length negotiation (set forth in OMB Circular AS-21).

G&P also reported that several RADAR grant expenditures for scientists Charlie mentored should be considered improper. For example, the largest amount the accountants deemed "improper" was a conference expense of $3,272.00 to Carla Pugh, MD and PhD, a young African American surgeon and academic, so that she could present RADAR findings. Indeed, she did considerable work for the RADAR project, the source of her payment, and continued an illustrious career.[5]

The accountants designated a similar travel refund amount that Charlie had asked the university pay to the American Society of Clinical Oncology national meeting for June M. McKoy, MD, JD, MBA, who is currently Associate Professor at Northwestern School of Medicine. She and Charlie Bennett were coauthors of many articles,[6] and she has remained deeply involved in RADAR. McKoy, originally from Jamaica, told me that working with Charlie Bennett "was the best thing that ever happened to me." She added, "He not only mentored me, he sponsored me; he saw to it that I had the opportunity to work with a distinguished physician." The G&P report did not mention her scientific work with Charlie or the reason for her official travel. The G&P investigation is replete with examples of bureaucrats rejecting university expenditures from NIH grants in support of MDs and PhDs who attended national conferences to present RADAR findings.

G&P decided that establishing a website was not a scientifically prudent investment. They questioned Charlie's choices of scientific collaborators. They characterized travel to scientific meetings to present RADAR findings as "inappropriate for payment based on standard accounting procedures." But no sensible person expects a scientist like Charles Bennett to base his decisions on "standard accounting." The more innovative Charlie was in his investigations, the more doubtful they seemed to Gould & Pakter.

James McGurk recognized the absurdity of the accountants' process.

> There are many, many aspects that are troubling, profoundly troubling about the government, or the OIG, in the investigation of Charlie. But the part that really gnaws me is the accountants saying, "We understand the science better than you understand the science, and therefore we are making the determination that what you spent it on was not appropriate. This was a waste of government assets." And on the other hand, we have people like Dr. Rosen, saying, "The results of his work were some profoundly significant articles that saved thousands of lives." Now who is it we should believe? The scientists or the accountants?
>
> I made the argument to Lindland, and he waved it off, saying, "I go with OIG. That's what I go with." Meaning, the OIG said these were the party animals; this was the funding of lifestyles of the rich and famous.

The long waiting period—roughly four years between complaint and settlement—was devastating to Charlie's career. Most people charged with murder or rape are given hearings more quickly than Charlie.

Shortly after the settlement with Northwestern, the president, provost, and dean of Northwestern University Feinberg School of Medicine publicly issued a joint statement, which needs a prominent place on the long, lamentable list of cowardly statements by academic officials.[7]

Message from Northwestern University President Morton Schapiro, Provost Dan Linzer and Feinberg School of Medicine Dean Eric Neilson Information and clarification regarding recent legal settlement:
August 2, 2013

EVANSTON, Ill.—We want to clarify recent reports and to correct inaccurate statements that appeared in the media regarding Northwestern University's settlement in connection with expenditures of federal research grant funds.

Northwestern University entered into a settlement agreement with the U.S. Department of Justice and the U.S. Department of Health and Human Services to end litigation over the expenditure of certain federal grant funds overseen by Dr. Charles Bennett, a former Northwestern faculty member who was the principal investigator on the grants involved in the settlement. The settlement involves no findings or admissions of wrongful conduct by Northwestern or any of its current faculty members or employees.

As we have made clear previously, Northwestern was nonetheless disappointed to see the allegations in the complaint because they are at odds with the University's commitment to a culture of compliance in the administration of federal research grants. Northwestern takes its grant administration responsibilities seriously, and fully cooperated with the government's investigation of these allegations in an effort to demonstrate their inconsistency with its institutional values.

Under the agreement, Northwestern will pay the government $2.93 million to resolve the covered conduct specifically identified in the settlement, all of which related to expenditures on Dr. Bennett's grants.

> [. . .]
> President Morton Schapiro
> Associate Dean for Science and Professor Daniel Linzer
> Provost Eric G. Neilson, MD, Vice President for Medical Affairs and Dean, Feinberg School of Medicine

The statement neatly skirts the border of accusation when it points out that the agreement deals with federal grant funds "overseen by Dr. Charles Bennett." The bureaucrats then announce that Northwestern is nonetheless disappointed to see the "allegations" in the complaint because they are at odds with the University's "commitment to a culture of compliance." The clear import is that Charlie administered federal funds in a manner "at odds with university policy and its culture of compliance." Curiously missing is that allegations are relevant only if they are proven.

In addition, by focusing on the grant funds "overseen by Dr. Charles Bennett," the officials absolve themselves—and "any of its **current faculty members or employees**"—from responsibility. That statement ignores the University's complex system for regulating grant expenditures: for Northwestern University funds and grants to pay Charlie's consultants and workers, *three* Northwestern administrators, a grant administrator, department chief, and a central-administration administrator *all* had to sign off. Where is mention of them in this statement? Rather, as written, the letter implies that Charlie himself was guilty of major misappropriation of grant funds, which cost millions of dollars along with heartache to the blameless leaders of Northwestern.

This official administrators' statement also fails to mention that its *former* employee, Feyifunmi Sangoleye, confessed to theft of the grants, that her guilt was known early on, that the university's layers of administrators neglected to oversee her false accounts. Rather, it mentions Dr.

Charles Bennett by name—the person whom the FBI, HHS, nor his own faculty had never interviewed.

These official administrators' public, published statement also neglects to mention its *own recent history of ineffective oversight.* Remember Schapiro's "the university's commitment to a culture of compliance in the administration of federal research grants"? In 2003 their university had been successfully sued by the US government for failure to oversee five federal NIH grants that were actually mismanaged by faculty members.[8] Surely these top administrators remembered their guilt and capitulation and $5.5 million fine: a former employee of Northwestern in the research department filed a qui tam action against Northwestern University, alleging Northwestern University committed fraud in the administration of its NIH grants. Northwestern quietly settled that one; the only notice of the qui tam faculty's name, Dr. Schwiderski, was in the Department of Justice announcement.[9]

In a 2006 academic workshop on "Enforcement of Federal Grant Accounting: The Legal Perspective," the moderator emphasized Northwestern's malfeasance:

> The gist of the [2003] allegations against Northwestern, as initially asserted by a qui tam whistleblower, was that Northwestern had included in its Institutional Base Salary charged to Federal grants the salary earned by clinical faculty from an independently incorporated clinical practice plan, but had failed to take clinical activity into account in its effort reporting system. It was also alleged that recipients of NIH career development awards had not in some cases dedicated to the awards the percentage of their total effort required by such awards (usually 75%), and that in some cases Northwestern had overstated the percentage of effort that other faculty were able to devote to Federal grants.[10]

The university chose to settle that FCA federal claim and refund *millions* to the government. No headlines, no accusatory letter published.[11]

The administrators' statement does not mention that neither the Department of Health and Human Services nor the Department of Justice found any evidence that Charlie had used grant funds for personal or nonscientific reasons. Any funds paid to any subcontractors were also paid by Northwestern University directly to any subcontractors. Northwestern University and the Robert Lurie Cancer Center submitted all requests for payment and obtained all funds directly. It was they, not Dr. Bennett, who had legal responsibility for developing, implementing, and overseeing all financial management systems, subcontracts, administrative compliance, and cost-accounting and budget protocols.[12]

Thus this official statement from Northwestern University should be recognized more for what its legal counsel *failed to include* that what it did include.

This official, misleading, and false statement remains a prominent posting in a Google search about Charlie, after all these years.

At every stage of the proceeding, Northwestern's administrators treated Charlie with suspicion and accusation. It started with the department's chief's supporting Camacho's claim that Charlie's use of his grant money violated the False Claims Act; was followed by Charlie's office being locked with no warning and for no significant reason; continued with the limits placed on his legal support and gulag-like conditions attached to his leave. It culminated in an official university statement that continues to besmirch his reputation.

It is difficult to avoid our conclusion that it was more important for Northwestern's trustees and administrators to continue their positive relations with major donors such as Amgen than it was to protect the rights of its faculty. For example, Ellen Kullman, former CEO of DuPont and a graduate of the Northwestern University Kellogg School of Management, has served on the Board of Trustees of Northwestern

University; and Shivani Gupta, 2008 graduate from Northwestern University's Biotechnology Master's Program and senior associate scientist at Amgen, is on the Industry Advisory Board of the Northwestern University Masters in Biotechnology Program.

To be clear: We did not uncover a smoking gun, but we kept smelling smoke.

Steve Rosen found the administration's actions appalling. "I don't think they were supportive of Charlie at all. They wanted to distance themselves from Charlie. I personally find it offensive." Rosen is of course familiar with the underbelly of academe:

> All this, and yet Charlie did nothing intentionally wrong. If we were to investigate all Northwestern *faculty* members for nepotism, for instance, you would find many guilty—wives, children—all on the payroll somewhere at Northwestern. So Charlie's hiring a family member to create a web page was no big deal—until the government made it so.
>
> The government wasted all that time and money, flying attorneys in from everywhere—for a miniscule amount that Charlie hadn't stolen or misapplied. Crazy.

A skillful, overall critique of the weakness of the case against Charlie and of the failure to include him in the settlement was posted on the *Health Care Renewal* blog[13] He wondered at the "naming and shaming" of Dr. Charlie Bennett for "relatively small alleged financial behavior" and explained to readers that "the institution that receives the grant is responsible for making all payments and disbursements related to that grant. The grant's Principal Investigator is responsible for the scientific conduct of the grant, but NOT payments, disbursements, business management, or accounting."

His blog concluded that "there are no allegations that any activities of Dr. Bennett or the university affected the quality of clinical research, clinical teaching, or patient care," which caused him to wonder why "Dr. Bennett's alleged misuse of grant funds required investigation by four different federal agencies, including the FBI." Dr. Poses's blog reviews a series of past settlements and reminds readers of the *institution*'s responsibilities when they handle grants:

> We have discussed numerous legal settlements by large health-care organizations, often involving hundreds of millions or even billions of dollars, sometimes involving guilty pleas by the companies involved or their subsidiaries. Almost never do these settlements name or involve in any way persons who authorized, directed, or implemented the misbehavior. In fact, we have commented again and again about the impunity of health care organizational managers and executives.
>
> Note that given the usual ways grants are administered, it would appear that Ms. Theis, the ostensible whistle-blower who will receive nearly one half of a million dollars from this settlement, actually may have had more direct responsibility for making the payments in question than did Dr. Bennett.

Dr. Poses's analysis points out that officials pursued the case against Charlie with unusual vigor, while they left the university and its policies/administrators alone. Without mentioning Amgen, he suggested its involvement.

> Dr. Bennett was the first author of a meta-analysis that was the first to show that epoetins increased the rates of adverse effects and death for cancer patients. This evidence lead to a reduced use of some very expensive drugs made by Johnson & Johnson and Amgen, and hence reduced revenues for both companies. . . . Dr. Bennett had financial relationships with Amgen. . . . He had reported serving [as a consultant] to Amgen. . . .

> Note that Amgen pled guilty to misbranding its epoetin, Aranesp, and therefore paid fine and civil settlement totaling $762 million. After writing that article [*JAMA* 2008;299:914–914], Dr. Bennett became known as a strong skeptic of the pharmaceutical industry.

He concluded, "Did Dr. Bennett's transformation from a consultant to a fierce scientific skeptic of the pharmaceutical industry influence the handling of this case, and its media coverage?"

No one's opinion seemed to matter to the university, an official blindness we cannot explain. After the civil settlement by Northwestern University for $2.9 million, Northwestern University officials told Stetler that they would no longer pay his attorney's fees because the case against the university had been settled. Attorney Howard Kline, Charlie's long-time friend, called David Rosenbloom and urged Rosenbloom to recommend that Northwestern University continue paying for counsel.[14] Rosenbloom not only summarily rejected his plea, but he told Kline that he was likely to urge the university to demand that Charlie return the money already spent on Stetler's legal fees.

Stetler asked for, and got, an additional month to try to work out a settlement for Charlie also. He tried a settlement, but Lindland continued to insist that Charlie pay a large fine ($1 million initially) and agree not to seek federal grants for his work—for the rest of his career.[15]

When he heard about the offer, Steve Rosen was infuriated: he had little respect for those government officials like Lindland, who insisted on payment for a noncrime:

> They may be straight, but they're just completely arrogant and ignorant of the academic world and what we do. And if they appreciated, here's a man who probably is responsible for saving tens of thousands of lives in different ways and had such a profound impact, and they're worried about whether he charged

> a bagel to a grant. I believe that's who they are. I think it was stupid; They're idiots.[16]

Stetler was able to convince Lindland to reduce his terms; after a few months, Lindland agreed to settle for $800,000 and a five-year, no-grants agreement between Charlie and the National Institutes of Health and other federal agencies.

At first, Charlie could not believe that such trivial issues, all of which he could explain if anyone ever asked him, should cost him so much money. He hadn't stolen any money or even used any inappropriately, so he refused to accept these terms. Stetler resigned—informing first Lindland, then the university, that he could no longer work with his client because Charlie refused to accept his advice.

James McGurk, subsequent attorney for Charlie, had no doubt that his client was innocent but bears some responsibility—Charlie was lousy at administrative detail and indifferent to offers of compromise. According to McGurk,

> If Charlie had said, "Okay, there's these bookkeeping requirements, and I didn't submit the receipts in the right order, and I didn't do this other stuff I was supposed to"—then you might have resolved the charges against him, and you would have left his enemies, at Northwestern or Amgen or even the bureaucratic opponents inside NIH, having little to say except, "He was guilty of jaywalking."
>
> There would have had something, but it would have been so insignificant, it would have been meaningless. Charlie Bennett was not prepared to admit even to Jaywalking. He felt that need to vindicate himself.

McGurk's conclusion was supported by one of Charlie's close friends, an attorney with considerable government experience: "Contrition is very helpful in achieving settlements with the government. Charlie doesn't do contrition very well."

For instance, Charlie might have acknowledged that he failed to follow prescribed procedures in hiring recent Swarthmore graduates as mentees. He had hired them directly to university positions for years, but, in 2008, he was informed that these mentees needed to be hired as temporary employees before they could be hired as permanent employees with full health benefits. "Yeah, Gould and Parker complained I didn't hire them appropriately according to the hiring regulations of the university. I just hired them."

When McGurk took over the negotiations in 2014, he was surprised by Lindland's hostility toward Charlie and his continued insistence that Charlie was involved in the ATSDATA fraud.

> This whole thing—I always keep going back to the fact that this isn't somebody in a laboratory looking through a microscope. . . . They were looking at this like, "You have a laboratory, and everyone is sitting at a lab and you get money to examine that, and anything else you do? You're wasting it."

Part of the underlying problem, McGurk theorized, was Charlie's unorthodox scientific methods: investigating international statistics and hiring outsiders to help. As McGurk became familiar with the intricacies of the case, he increasingly focused on the threat Charlie posed to Amgen's financial entanglement with the private university Northwestern.

> He was the person who would say, "The Emperor has no clothes" when everyone else was fawning. "Oh, my god! We need to fund our program; these sponsors are so critical." Northwestern is so heavily involved in terms of members of the board. It's that the tentacles of big pharmacy are so much more deeply into Northwestern than they are into almost any other major medical research institution.

Lindland was in a strong bargaining position because he could delay settlement, and Charlie would continue to be hurt professionally by the pretrial accusation. Lindland made clear both to Charlie and to his attorneys that if Charlie did not compromise, he would face a long, determined campaign marked by continuous accusations through public statements from the US Attorney's Office restating his guilt.

The threat was real. As we have already shown, academics live and die by their reputations, and any continued public attack would surely damage and probably kill any hope Charlie could have of resuming his research. Charlie was not sure how to respond to this threat. Charlie really, really wanted his day in court. He spent his spare time creating arguments, strong language, counterarguments. Yet, attorney friends uniformly advised him to seek a settlement-oriented attorney. "You will never be able to force an assistant US attorney to go to trial on an issue like this within five years," one told him, pointing out that the AUSA were, to a person, experts in using negotiations to force concessions.

Along with other attorneys, McGurk pointed out that continuing to insist on his innocence would cost him years and many hundreds of thousands of dollars: better to accept a settlement if he could get one that he could afford.

Stetler urged Charlie to accept Lindland's revised proposal, arguing that the process might last for years—plus Charlie might lose at a trial, even if Lindland could produce only minimum evidence of misconduct. Charlie refused.

Serious negotiations began in 2013. Lindland continued to insist that he would agree to settle the case only if Charlie paid $800,000 and would agree not to apply for or be funded by any federal grant—for six years. This was a difficult negotiation.[17] At one point, Charlie was informed that Lindland sent one of Charlie's lawyers a picture of Charlie's house, along with a reminder that it could be lost in the battle. Remember: Google maps did not begin photographing homes in 2007. It would be obvious that if the Bennetts were forced to pay Lindland's demands, Amy's house sale would be part of the money they would have to send the government. Illinois law protects homeowners during bankruptcy, but not federal legal claims.[18] Constant negative headlines could impact Amy's job futures. The whole episode reminded Charlie of the *Godfather* movie, where a gangster who had betrayed the mob awoke to a severed horse head bleeding on his bed.

Negotiations remained deadlocked for almost a year. Sometime in September 2014, the first-ever face-to-face meeting between Dr. Bennett and his counsel McGurk; Linda Wyetzner, counsel for Melissa Theis, relator; and Lindland took place.[19] Charlie recalls that Lindland, like most AUSAs, knew how to pressure an opponent. According to McGurk, Lindland repeatedly stressed the strength of his bargaining position: "Look. Did you see the press release I wrote about you? I can put out a press release like that every month. . . . I'm going to have your job in six months with six press releases."

Lindland reminded Charlie of the government's ability to mold the facts. He hadn't, for instance, put any mention of Fayi in the press release. She had already pleaded guilty—and yet the government attorneys had never mentioned her name or role in the theft to the public. He had not broadcast Feyi's penalties. He said, "I can write whatever I want. Your job is gone."

McGurk was surprised by Lindland's continued belief that Charlie was involved with Fayi's theft when he must have known that Charlie could have worked with Amgen and made much more money:

> He was this true believer in the stupid Fayi business. But I think it's totally stupid. I mean this is chump change. Everything about the Fayi business was silly on that aspect. And combine that with the headhunter view of the Inspector General people saying "We got to; we're going after people no matter how prominent. You got to follow the rules. We don't care if you're on the short list for the Nobel Prize."

In our one phone conversation, Lindland made clear that Charlie's eminence made him enticing as a defendant. McGurk was aware of this motivation and believes that Charlie hardheadedly played right into it:

> They wanted to be in the newspaper and scare the excrement out of folks so that everybody would comply with their obligation to file tax returns. "They got that actor! They got that judge! That doctor!" That's what they wanted. They wanted the biggest possible name. And in that regard, Charlie fell into the trap of stressing his eminence and importance.
>
> And among the many, many features of Charlie Bennett, humility is not one of them.

Charlie's legal fees were more than $10,000 a week for the fourteen months before he finally agreed to settle. And not surprisingly, the constant drumbeat of allegations and misstatements made Charlie fear that his career, previously marked by constant success and national recognition, was in jeopardy.

After months of stalemate, Lindland announced that he was "feeling charitable." He made what he described as a "one-time offer" to settle for $475,000 plus interest. McGurk considered both the strength and threats of the government—and also Charlie's future: "This is the best you can get." Charlie, exhausted and confused by the relentless attacks and battling, agreed to pay the government the $475,000. A legal—but galling—17 percent of that was to be paid to Melissa Theis. There was no happy ending for Charlie, who was forced to mortgage his home in South Carolina, and, with both shame and fury, had to borrow $100,000 from friends and family to pay the agreed amount plus the $600,000 legal fees.[20] Lindland generously informed Charlie that if he could not find the money, then Lindland would be happy to accept Charlie's retirement accounts to complete the payment.

Moreover, the settlement agreement with Lindland specifically left open the possibility of future government action that would adversely affect Charlie's ability to continue his career.

Section 5 of the settlement specified that, "Nothing in this Agreement precludes any agency in HHS from imposing corrective actions or special award conditions on any grant including, but not limited to, future grant awards."[21]

Charlie received notice that his medical license would be revoked; he learned that the HHS were going to place a lifetime ban on his grant ability. In 2014, McGurk was informed—with no explanation—that Health and Human Services officials were indeed planning to permanently disqualify Charlie from applying for, or participating in, any grants that were administered by any federal agency. McGurk traveled to Washington, D.C., to argue on his client's behalf. He sat across the table from thirteen government attorneys. McGurk presented them with a four-page summary of Bennett's issues, beginning with the sealed Theis complaint. He provided them with the full picture:[22]

- No one has ever in this case challenged any of Dr. Bennett's scientific work.
- The University of South Carolina reviewed all of Bennett's work at USC and found the accounting correct.
- Bennet has supported the US government claims in another False Claims Act matter.
- Bennett is currently providing information on adverse drug effects to government counsel in ongoing investigations.
- Dr. Bennett's work has saved thousands of lives.
- Dr. Bennett does not accept pharmaceutical funding.

McGurk pointed out that the HHS proposed prohibitions would prevent Charlie from replicating the scientific successes he had achieved previously by virtue of government grants.

The government lawyers responded in two ways that were mutually supportive, although essentially inconsistent. "We all have to play by the same rules and being an important scientist does not give him the right to commit fraud," they argued. Beyond that, they claimed his reputation was overrated and even called him "Champagne Charlie."

"Champagne Charlie" indeed. It is difficult to find a title less fitting for Charlie Bennett, a person who viewed each new grant as an opportunity to do more demanding work. "Champagne Charlie," for no monetary gain, mentored generations of aspiring scientists. We found that he volunteered, without pay or any recompense, to advise and testify for patients harmed by Cipro and Levaquin. How did the HHS come to see Dr. Bennett in a negative light? The government's functionaries' erroneous vision of Charlie Bennett was evident to some of his colleagues and mentees who were interviewed by the FBI.

Dr. Elizabeth Calhoun, for instance, recalls that she was so annoyed by her FBI questioners' vision of Dr. Bennett that she finally asked the FBI agent, "Do you have any idea of who it is that you are talking about?"[23]

Those who have worked with Charlie regularly use the term "workaholic" to describe him. Benjamin Djulbecovic[24] told me that if Charlie were to be walking across the nicest, most beautiful beach in the world with perfect sunshine, "He wouldn't notice; he'd be thinking about science." David Meyer, CEO of Eastern Research Group, a leading consultant for the FDA, has known Charlie since their Swarthmore days. His image of his longtime friend is also quite different from that of the government lawyers and accountants: "Charlie was at the top academically, and he was also the kind of guy that would take time to help other students who were having a hard time, which was a lot of people. It was just out of the kindness of his heart."[25]

McGurk was surprised to hear the government attorneys argue that Charlie's work was essentially derivative and not very important. Steve Rosen emphatically disagrees: "Charlie was one of the most creative investigators. He's responsible for unraveling these rare adverse events with many different compounds, and then bringing it to public attention."

McGurk could not stop the nightmare. After Charlie voluntarily reassigned two of his SONAR grants for a few months (each time to other members of the faculty) he unexpectedly received a letter from Amy Haseltine, Suspension and Debarment Official and deputy assistant secretary of the Department of Health & Human Resources.[26] This announcement would be published in the General Services Administrator's system of contractors "debarred, suspended, proposed for debarment, or declared ineligible by any federal agency." The terms of the debarment were quite broad; not only was he debarred from contracting with the government, but HHS instructed him that he could not act as "an agent or representative of" other entities dealing with the government. Three times (2012, 2013, 2014) HHS removed him from his own research grants; Northwestern gave them to other faculty members.

Indeed, Charlie was now "prohibited from participating in certain non-procurement transactions which include but are not limited to grants, cooperative agreements, scholarships, fellowships, contracts of assistance, loans . . . donation agreements." In practice, he was told he might "not meet with or talk to anyone involved in any of his previous grants."

As a result of the debarment, Charlie was dismissed as a participant from a variety of programs directed at adverse drug reactions. The most significant of these was a developing program to counter opioid usage, headed by Professor Mary Lee Larson of Brandeis University. The day after he received Hazeltine's letter of debarment, Larson called and removed him from participating in an important program that he had helped to inaugurate.

Faced with professional disaster once again, Charlie decided he needed new counsel with different experiences. He contacted a highly regarded former assistant US attorney, Rob Henning, who had taken on Amgen in a False Claims litigation on behalf of Amgen employees.

While Henning was deciding whether to take Charlie's case, Amy Bennett wrote to him expressing her fear for her husband's future:

> My fear is Charlie is not going to survive this. He has threatened suicide many times. . . . Please call him and give him some hope, if you can. As you know, he is waiting to hear if he gets his appointment continued here at USC. That is still up in the air.[27]

Fortunately, Henning decided that he could not leave Charlie adrift. He took the case and negotiated an agreement with HHS's Haseltine that Charlie's disqualification would be eventually lifted, in return for Bennett's taking forty hours of classes in grants management. We at first thought this was a Charlie joke. Nope. Charlie was to take forty hours of classes rather than a professional disqualification.

Thus, from roughly 2009, legal problems made it almost impossible for Charlie to obtain grants for the type of work he had so successfully completed during his days at Northwestern. In particular, the US Attorney's Office for the Northern District of Illinois and Assistant US Attorney Kurt Lindland acted as though it were in the public interest to block Charlie's scientific work. The enmity was relentless. As Charlie describes, "it takes me forever to get a job. Then in 2013, Northwestern settles and carves me out of the settlement and reinitiates everything. In 2014, the government actually makes me pay them. In 2015, they suspend my grants again. In 2015, they try to take away my license. In 2016, they try to suspend me from the research."

By 2016, when Charlie was legally allowed to resume his work, he was broke and in debt because of legal fees and settlement costs; the NIH had severed all of his earlier collaborative agreements.

The emotional, financial, and professional costs to Charlie Bennett from these accusations and procedures were enormous. Charlie has little doubt that the government's relentless focus on him was based on pressure initiated by Amgen.

CHAPTER 9

Big Pharma Gambles on Thalidomide: From Medical Disaster to Wonder Drug

In 2016, Beverly Brown, an employee, took on pharmaceutical giant Celgene for its unlawful and unethical promotion of the renewed use of thalidomide. Her lawyers got Charlie, as an expert witness, to help expose the unlawful practices.

Like EPO and Flox, thalidomide had considerable value but required government supervision. A drug with a troubling history, thalidomide was first developed in Europe around 1958. Because tests showed that it had sedative effects,[1] international doctors often prescribed it for morning sickness. The result was tragic. We all can remember accounts of thousands of mothers treated with thalidomide who gave birth to children with serious birth defects. Many of the children had severe limb malformations: children with small hands attached to their elbows. Others were born with congenital defects affecting ears, eyes, heart, and kidney. After public outcry and a courageous stand by another pharmaceutical hero, FDA head Dr. Francis Kelsey, in 1961 thalidomide was withdrawn from the market to international relief.[2]

But only a few years later, in 1964, research scientists discovered a unique therapeutic use of thalidomide.[3] Patients with leprosy, given the drug because of its sedative effect, experienced improvement of their lesions. Twenty-five years later, the FDA granted marketing approval to Celgene for thalidomide to treat cutaneous manifestations of ENL (a form of leprosy).

Meanwhile, in 1997 cancer research scientists began conducting treatment trials to evaluate thalidomide and at the University of Arkansas discovered the unexpected additional benefits identified in patients with multiple myeloma.[4] Simultaneously, the FDA gave Celgene permission to market Thalomid and Revlimid, two advanced forms of thalidomide that could be used to treat several forms of cancer. The once-despised drug suddenly became a standout in the development of treatments for cancer.

The FDA was cautious: because of thalidomide's dangers and past history, the FDA required Celgene to develop a Risk Evaluation and Mitigation Service (REMS) program to instruct physicians on its proper and limited use for cancer. Celgene developed its program and issued a variety of statements attesting to its safety and effectiveness.

The FDA had not been cautious enough. Celgene used the REMS program as a marketing device and hired a large task force of salespeople to whom it gave some training, along with misleading titles such as "immunology specialist" or "hematology consultant." These mislabeled Celgene employees contacted physicians—ostensibly to explain the REMS program, but actually to focus on the wide advantages of the product and sometimes to enlist the physician to participate in its widespread program of advertising.[5] In addition, Celgene representatives offered the physicians a chance to participate in "ghostwriting" articles[6] about the benefits of Thalomid and Revlimid.

In a 2006 collaboration with Senator Richard Blumenthal (then-Connecticut attorney general), Charlie filed a citizen petition requesting that the FDA require Celgene, which was producing Thalomid, to warn physicians and users of the risk of venous thromboembolism (VTE) in *off-label settings.*

They launched the Bennett-Blumenthal petition[7] after investigations revealed that 92 percent of thalidomide prescriptions were for off-label use.[8]

The FDA approved their petition, and Celgene agreed to add VTE warnings to its labels.

Ten years later, Celgene employee Beverly Brown, as relator on behalf of the United States and twenty individual states, filed a qui tam suit[9] against Celgene Corporation. The complaint alleged that Celgene "unlawfully marketed blockbuster drugs Revlimid and Thalomid accounting for more than $1.7 billion and $440 million respectively."[10]

Beverly Brown had worked for Celgene for almost ten years as an "immunology specialist" before filing suit. Her suspicions began when she was instructed to alter codes that were to be applied by physicians to make drug prescriptions reimbursable. She contacted the FDA, which, after investigating, confirmed her suspicions.[11]

The unlawful marketing technique charged to Celgene includes almost all those discussed earlier with respect to EPO and Flox, including "marketing its drugs for off label use,"[12] "paying kickbacks to induce drug sales," and "making misrepresentations with regard to the safety and efficacy." Brown's legal team also accused Celgene of "directing and controlling physician speaker programs that purport to be unbiased."

During the discovery process, Brown's attorney Justin Brooks also learned that a major technique of Celgene was the widespread ghostwriting of favorable scholarly articles used in the promotion of Celgene's drugs. According to Brooks,

> Celgene actually had a contract with Excerpta Medica where they would basically provide these drafts and they would go to their marketing department, and they would basically insert messages that they wanted and edit it, then it would go back to the authors. . . . We argued that it was a violation of the Food, Drugs, and Cosmetics Act. Because they were taking these articles about off-label uses and directing what they said, how they said it, and they were arming their sales representatives with these, supposedly, independent publications they call "editorial assistance."[13]

In the Undisputed Statement of Facts were lists of company executives and "strategic contactors" (doctors) who prepared articles for which the "author" was paid money, honoraria, and appointed to advisory committees.

Celgene's lawyers denied that the alleged facts constituted a False Claims violation. They admitted off-label marketing but argued that the drugs in question are "two life-extending drugs, Thalomid and Revlimid, that physicians across the country prescribe to treat some of their most seriously ill cancer patients." And it claimed that off-label marketing did not meet the standard of "knowing falsity" required by the False Claims Act. Thus, it moved to dismiss for failure to state a valid claim.[14] Celgene provided testimony from an impressive list of eminent oncologists in support of its conclusion.

Attorney Brooks had a difficult time finding an expert to counter Celgene's medical claims. He immediately thought of Charlie Bennett but was initially worried because of the highly publicized false claims charges against him. But, finally, he decided that he had no viable choice: most of the experts in the field were financially involved with

Celgene. And he did not believe that the False Claims Act case against Charlie was legitimate. "I firmly believe that it was Amgen behind that."

Importantly, Charlie's written statement in support of the plaintiff did not question the effectiveness of the drugs; rather, he attacked Celgene for a variety of unethical behaviors.[15] In particular, Charlie pointed out that Celgene did not adequately disclose *dangers that it knew about.*

> In 2013, following the culmination of years of study, national drug safety expert Tom Moore and I reported that *only 1% of all VTEs/DVTs associated with thalidomide were ever reported to the FDA.* Now, after reviewing the documents produced by Celgene in this case, I can appreciate why. Those documents indicate that Celgene avoided documenting serious adverse events and suppressed evidence of adverse events that were not reported to the FDA to avoid regulatory scrutiny. [emphasis added]
>
> For example, I reviewed an October 2000 email exchange from [field rep worried about adverse experiences]. . . . Celgene's Chief Medical Officer responded, "Please do not send such explicit e-mails to the field. This is a non-erasable message that can be audited by the FDA."

Only 1 percent of all serious adverse reactions were reported to the FDA.

Charlie also went after Celgene for off-label advertising through ghostwritten articles.

> For years, articles on off-label (and on-label) uses of Celgene's thalidomide drugs have appeared in journals under the putative authorship of oncologists and hematologists. [However], disclosures of "editorial support" of Excerpta Medica and "financial support" of Celgene do not come close to identifying the nature

> of Celgene's relationship with Excerpta Medica, Celgene's role in the planning of these publications, Celgene's ability to dictate or influence publication content, and the role that these publications play in Celgene's larger promotional strategies.
>
> Celgene's documents further show that, in many cases, Celgene and Excerpta Medica already had a specific manuscript planned but had not yet chosen an author or target journal. . . . Although authors reviewed the outline, drafts were then reviewed by Celgene as well as the putative authors. Celgene had to approve the final draft in order for submission to journals to take place.
>
> Neither scientists nor academic authors acknowledge this shady behavior; while not illegal, it turns "scientific professional" into "scientific staff."[16]

After some legal skirmishing, Charlie's statement was released to the press. This damning statement was, of course, met with strong resistance. In response, Celgene immediately scheduled a deposition for Charlie. At the deposition, Charlie made clear that he was not faulting the physicians but instead was focusing on Celgene—for not being clear about the risks posed by its product.

At the end of Charlie's deposition about the pharmaceutical aspects of the case, the cross-examination by Celgene's lawyers became personal and hostile—questioning his honesty and professional standards based on the earlier False Claims Complaint against him. They quoted from statements made by Lindland and Northwestern officials. They asked if he had enjoyed his vacations on government money; they asked where he went and how much they cost. They asked him to explain how he and Feyi had managed to obtain $80,000 of government money for a nonexistent vendor.

Brown's lawyers did not object to or dispute the legitimacy of the deposition questions. Charlie recalls only once being instructed not to answer a question. They afterward explained to a shaken Charlie that it is customary to allow great leeway in depositions. Furthermore, he was not their client, so any rigorous objection from them would be inappropriate.

When the hearing ended, Charlie was shaken. "I felt like a piñata set up as a target and getting hit from all sides."

Celgene lawyers contested the premise that off-label marketing constituted a false claim. They argued that in cases such as this, off-label marketing did not meet the standard of "knowing falsity" required by the False Claims Act. Celgene moved to dismiss for failure to state a valid claim.[17]

We agree with Celgene: there is force to these arguments. The False Claims Act was meant to outlaw efforts to obtain money from the government based on knowingly false or fraudulent representations. It takes significant stretching of the language and departure from its purposes to hold that the language applies to off-label marketing based on positive medical evidence.

Beverly Brown's attorney Brooks sought to meet the "knowingly false" standard by arguing, "Celgene is liable under the FCA because it engaged in a systematic campaign to encourage doctors to write off-label prescriptions of Thalomid and Revlimid when it knew these prescriptions were not reimbursable under the relevant government programs but would nonetheless be submitted for reimbursement."[18]

At the request of Brooks, Charlie prepared a rebuttal statement:

> All of the expert reports suffered from the same deficiency: either *no* consideration of the safety aspects of distribution and use of Thalomid and Revlimid (particularly off-label use) or wholly inadequate assessment of these safety concerns.

Judge George King accepted the plaintiffs' theory that claims submitted to Medicare seeking reimbursement for off-label uses of Thalomid and

Revlimid were false "because cases in this circuit and elsewhere holding that a claim is 'false' if it is statutorily ineligible for reimbursement," which these items were.[19]

Judge King's decision seems consistent with the weight of authority as applied in similar cases. Thus, almost automatically, off-label marketing will be vulnerable to assertions of false claims. Whether or not this is sound public policy is not clear—at least in situations like the Brown case where the drugs in question had already won the support of leading clinicians.

After Judge King's ruling, the case was set for trial. Not surprisingly, Celgene was eager to avoid a trial that would reveal both its off-label marketing efforts and its widespread use of ghostwritten articles favorable to its products. Celgene thus offered to settle the case by a payment of roughly $300 million to Brown and the various state plaintiffs.[20] Its offer was accepted.

This case ultimately demonstrates both the importance of pharmaceutical innovation and the need for constant surveillance of their methodology. Thalimid and Revlimid each have great potential and were well worth developing. Since the hearing, they have become more widely acceptable. Revlimid, in particular, is now a well-accepted drug.

But the methodology used by Celgene was reprehensible—from the phony characterizations of its salespeople as "experts," to the ghostwritten articles, and to its documented failure to report on adverse reactions. Celgene did not pay adequate attention to the safety of its customers. Charlie's role in holding them to public and legal account was vital.

Because of the well-publicized charges against him, Charlie was not the perfect expert witness. Brooks, however, found him admirable: "I think that he saw something that he thought was wrong. He is legitimately concerned about drug safety. And he decided to take this on, wasn't a popular position, and wasn't going to earn him much of anything, other than what he thought was the right thing to do."[21]

CHAPTER 10

How Great Institutions Affect Academics and Scientists

Charlie's story offers interesting, sometimes confusing, sometimes contradictory insights into the pharmaceutical industry and its influence on both medical academia and the justice system. This is not a new story,[1] but it is a new way of looking at the harms created by these institutions.

Big Pharma and Medical Academia

This is an age of opportunity, notoriety, and wealth for Big Pharma. For a variety of reasons, including increased computer capabilities along with increased scientific understanding, pharmaceutical institutions are able to develop powerful new products and significantly improve existing ones. Because of these great advances and a variety of societal changes including the aging of the population, pharmaceutical companies are able to make unparalleled contributions to the health of the public and unparalleled profits.

We have included throughout the book several stories that have, at their core, lasting scientific achievements; e.g., Goldwasser's ultimately successful twenty-year struggle funded by Amgen to develop a drug capable of inducing the body to produce additional erythropoietin.[2] We trace this achievement through Bayer's subsequent discovery of ciprofloxacin—still recognized as a major medical achievement.

During the time period covered by this book, Amgen was transformed from a struggling start-up to a pharmaceutical giant. Amgen uses this story for self-promotion—their official story ignores, with virtually no mention of, Goldwasser. This book reveals that, in pursuit of growth and profit, Amgen officials played fast and loose with the truth. They consistently denied the validity of Bennett's carefully conducted study that demonstrated the dangers of erythropoietin. Years later, investigators uncovered Amgen emails discussing the need to refute his allegations.

We know that in 2007 Amgen officials were concerned with Charlie's analysis of the dangers of ESAs. They learned of his evolving position through the *Cancer Letter* article anticipating his poster presentation at ASCO; the article repeated Charlie's conclusion that "the use of erythropoietin stimulating agents in oncology is associated with statistically significant increases in venous thromboembolism and mortality."[3]

We know that it was this published conclusion that led Amgen's vice president Baynes to attend the ASCO meeting and to confront Charlie before his presentation—a confrontation repeated to us by Charlie that ended in the threat we described in chapter 3.

> He said, "Bennett. I read your submission draft that you sent to the *Journal of Clinical Oncology* with the 2003 cutoff point."

> And I said, "What do you think I should do with the draft?" And he said, "Here's the waste basket. That article is garbage and goes in that wastebasket."
>
> I said, "Well, actually, I was asking about where I should publish it." But he said, "If you publish this, I will personally destroy you."[4] I changed my speech after our meeting to get rid of the 2003 timeline. He really scared me.

A short while later, a US attorney for Western Washington, Peter Winn, enlisted Charlie for an investigation of Amgen. The US Attorney General's office had recovered Amgen emails during their False Claims Act litigation against Amgen. In one of the emails, the US Attorney's Office learned more detail about that ASCO presentation.[5]

For example, Amgen executive director Trish Hawken had "set up" Dr. John Glaspy of UCLA—one of their strong supporters and grant recipients—to respond to Charlie's ASCO presentation. In the uncovered memo, the US attorneys learned that Glaspy wrote back to her after he sat in on Charlie's presentation to say that he couldn't challenge something that wasn't discussed. "So far, no calls. Also, he did not rise to the bait I set out Saturday evening. . . . (I was waiting with a mallet, complete with backup slides to bash him). . . . He may be smarter than I thought."[6] Amgen's chief of research Pearlmutter was no fan of Glaspy's: "We need "to provide third party experts who—unlike Glaspy—can stay on message."[7]

The uncovered emails reveal a corporate effort to negate his scientific analysis. On February 26, 2008, Kelley Davenport, Director of Amgen's Corporate Communications, sent an email to "14 Amgen persons" of the forthcoming "Press Release re: Charlie Bennett's meta-analysis of ESAs":

> Unfortunately—even though the meta-analysis is old news, there is quite a bit of media interest. We have arranged for

> Roger P. [Amgen Chief of Research] to speak to the following media outlets this morning to make sure our message is included in the media coverage. *Dow Jones, Bloomberg, NY Times, USA Today.*[8]

Again, we are not stating that we found the smoking gun. But there's sure a lot of smoke coming from these internal emails.

Amgen also learned that a powerful Congressman was becoming critical of its activities: based on Charlie's findings, on March 6, 2008, David Brier, senior vice president for global affairs of Amgen, sent an email to top Amgen officials[9] expressing concern about a letter highly critical of the way the company marketed EPOs written by Congressman John Dingell, chairman of the House Energy and Commerce Committee, and Bart Stupak, chair of the Oversight Committee to the FDA.

VP Brier was obviously worried. Brier wrote that he had seen Dingell's FDA letter, and that within the letter Chairman Dingell and Subcommittee Stupak claimed:

> 1. Companies have "spent millions of dollars lobbying and marketing [on] marketing campaigns that misrepresent the risks with ESAs."
> 2. The "dramatic findings" of the Bennett JAMA article show a 57% increase in the risk of blood clots."
> 3. The Bennett analysis "discusses results of experiments that indicate ESAs could actually *enhance* [italics in the original] tumor growth."
> 4. 10 clinical studies have been halted by Data Safety Boards for ESAs.
>
> . . . this letter will be widely reported and will have some effect on the FDA. Our challenge is determining how to effectively respond to the conclusions about potentially pulling the indication, as well as their allegations.

Brier is concerned about Dingell's letter, in which he implores the committee to ask that Amgen "reconsider the risk/benefit profile for this potentially dangerous class of drugs." Dingell has asked that the committee discover "whether any demonstrable benefits of these drugs offset the evidence of increased mortality, blood clots, and tumor promotion." Lastly, Brier confesses,

> the letter notes that the FDA should act with "due haste" to avoid further endangering patients who may have been steered toward these drugs by aggressive marketing practices, including the use of misleading direct-to-consumer [ads] and recommendations from physicians who personally profit from the sale of these drugs.

Clearly, Brier warns, "given the committee's position, history, and jurisdiction, this letter will be widely reported and will have some effect on the FDA. Our challenge is determining how to effectively respond to the conclusions about potentially pulling the indication, as well as their allegations."

Amgen was on full alert.

Attorney James McGurk, himself a former AUSA, extensively investigated the files and interviews and came to believe that Amgen played a major role in creating the case against Charlie:

> I saw communications, emails and internal communications, in Amgen saying, "We gotta get this guy. This guy is a risk. This guy is a problem. We have to target him." And of course, it's done on a super-sophisticated level. It's not like we have our press department issue a press release saying, "This guy's full of . . ." No, no, no, we get a friendly professor at a school.
>
> It's clearly, they have a focus and they're sophisticated to know we can't disclose any of this, this has to be on the deepest

> of background. He was up against a very, very wealthy, sophisticated opponent in Amgen. He was up against, what do I call, very righteous, institutionally arrogant folks at OIG. He was up against personally vindictive folks at Northwestern in terms of the campus politics side. I mean, this guy had some enemies. He really did.[10]

Amgen's eagerness to shape scientific literature is further demonstrated by the story of young, up-and-coming medical researcher Dr. Marshall Horwitz and Dr. David C. Dale, a prominent academic and the former dean of Washington University's medical school.

From 1998 to 2000, Horwitz and Dale worked together in an effort to find a gene responsible for a hereditary form of the blood disorder neutropenia.[11] The collaboration ended with differences about the future direction of research and Horwitz's "discomfort" with Dale's strong endorsement of Amgen's products. By comparing Dale's articles with that of others, Horwitz grew suspicious that Amgen ghostwrote journal articles that were attributed to Dale[12] that repeatedly praised Amgen's products. Dr. Dale denied that there was a problem.

Horwitz was concerned that Dale was promoting Amgen's products; he formally charged Dale with academic misconduct. He submitted written evidence of Dale's improper collaboration with an industry-sponsored writing firm, Gardiner-Caldwell Communications (GCC).[13] Sue Clausen, a vice president of the University of Washington School of Medicine, investigated Horwitz's claim and reviewed a confidential email sent to Dale from GCC's Andrea Cole. Dale had written GCC advising, "The manuscript is excellent. I approve and please proceed to send to editor and publisher. Thank you." The European *Journal of Gynaecological Oncology* published the article,[14] with Dr. Dale as the sole author—with no reference to the contributions of GCG or Amgen and with the order of his names mistakenly scrambled.

VP Clausen found no plagiarism or academic abuse. Dale then returned to the article and published an erratum, stating that he received help from Amgen and that the principal ghostwriter Jackie Williams (without naming her) was inadvertently omitted from the final draft.

It was a shady enterprise at best.

The failure to acknowledge the work of GCC and Amgen in Dale's article was surely not accidental.[15] Amgen designed the article to present the appearance that it was a report from the frontiers of science on the value of Amgen's products but, in the hands of Amgen salesmen, the article might well convince physicians and journalists of the value of Amgen's products Aranesp and Neulasta. Dissatisfied with the academic response to the Dale/Amgen misconduct, Horwitz contacted the US Attorney's Office in Seattle and met with the assistant US attorney for Seattle, Peter Winn, who studied the evidence and agreed to both join a qui tam case against both Amgen and Dale. Winn started issuing subpoenas and requesting information. When the production rolled in, Winn employed Horwitz to review all the documents.

The material that Horwitz examined for Winn revealed, among other things, that Amgen had developed successful techniques for seducing academics and physicians. Horwitz recounts that Amgen's messages in the emails and letters were filled with flattery and went something like this: "We are contacting you because we are impressed with your work," "We consider you an opinion leader, a practice leader," "We'd be happy to fund a study. We'd like you to be on our speaker roster."[16] Many pseudoscientific articles followed, as Horwitz explained:

> It was clear that ghostwriting was going on. . . . It didn't take anything very sophisticated to figure out because they called it "publication planning."
>
> Publication was a part of the marketing strategy, and they plan what papers they're going to publish, who was going to write them, who was really going to write them, who they

> thought would be a good sham author to have. You know, first, second, third choice of all authors.
>
> And there were dozens of already-written papers where the byline, the authorship said, "to be determined." And they had discussions about who would be "authors."

Winn filed *U.S. ex rel Horwitz* in Seattle in February 2007; the defendants were Amgen, David Dale, and GCC as the ghostwriting enterprise. Winn's suit came as a companion to a larger, collective suit against Amgen.

Five years later, the government announced that Amgen would pay $762 million in fines and payments to settle the grouping of false-claims accusations by multiple US attorneys general: "Amgen Inc. Pleads Guilty to Federal Charge in Brooklyn, NY.; Pays $762 Million to Resolve Criminal Liability and False Claims Act Allegation."[17] It was one of the largest false- claims settlements ever, with the most money ever paid to the government and relators by a drug company, a company that had to admit publicly that it sold drugs that it encouraged physicians to prescribe for off-label use. The federal criminal settlement was highly publicized.

While the ghostwriting issue was of great potential interest, Amgen had insisted that, in return for the $762 million settlement, ghostwriting would not be raised in Department of Justice post-settlement comments. The department agreed.[18]

The settlement ended the case against Dale, who was a named defendant, without a trial or verdict. Neither the US Attorney's Office nor the university took further action. Dale remains on the faculty.[19] Horwitz has found it difficult to recapture the intellectual momentum that once seemed so natural for him.

It is difficult to measure the costs and benefits of the case for Horwitz. On the plus side, as a qui tam relator, he received a significant payout—more than he would have earned in salary in ten years. On the negative side, he remains in a state of academic isolation that continues to this day.

Ghostwriting is only one of many techniques pharmaceutical companies use to deceive doctors and the public.[20] The pharmaceutical companies also fail to publicly acknowledge significant flaws in their products.[21] For example, for many years, both Bayer and Johnson & Johnson refused to acknowledge the devastating side effects of ciprofloxacin.

Even more deplorably, they launched attacks on critics: we wish that medical researcher Horwitz, whistleblower Brown, and Charlie's experiences with the retributions of Big Pharma were unique.[22] However, as the following examples show, that is far from the case:

* * *

In 1995–1996, Dr. Nancy Olivieri, an internationally renowned expert on blood disorders, prepared to alert the public to her findings that Deferiprone might actually be toxic.[23] She identified an unexpected risk in industry-sponsored clinical trials[24] involving patients with thalassemia—an inherited, potentially fatal blood disorder. When she moved to inform patients, the company issued warnings of legal action. The drug's manufacturer Apotex Inc. prematurely terminated drug trial results for the potentially toxic drug. Nevertheless, Olivieri informed her patients and the scientific community of the risks she had identified. Dr. Olivieri published her findings on the drug's risks in a leading scientific journal in 1998. She was removed from her faculty position[25] (but later reinstated after prolonged publicity). When Apotex later approached the FDA for approval, they excluded her data. As Dr. Olivieri has written, "In licensing aducanumab, the FDA ignored the advice of its own independent advisory committee. To date, three of the committee members have resigned in protest."[26]

Dr. Gideon Koren, a former Toronto University collaborator of Dr. Olivieri, remained a strong supporter of the drug and of Apotex. How strong? He created and distributed a series of anonymous poison-pen letters to her colleagues and media disparaging her personally and professionally, referring to her colleagues as "a group of pigs."[27] Koren, a Hospital for Sick Children's physician, for over eight months denied responsibility for the hate mail during a public-funded investigation commissioned by his employers, Hospital for Sick Children. Then, Dr. Olivieri's colleagues obtained DNA from a stamp Koren had inadvertently licked to send the hate mail, and that bit of science ultimately revealed his authorship.

There is no indication that Apotex or the university had anything to do with Koren's emails. While all this was happening, though, the University of Toronto and Apotex were in negotiations for a significant donation.[28]

When the Hospital for Sick Children reviewed all of Dr. Koren's published work, they found problems in about four hundred of his published articles. Finally, the then–Dean of Medicine at Toronto University withdrew his support of Koren after Koren was found guilty of research misconduct. The scholastic misconduct forced Koren to relinquish his medical license. He moved to Israel. We did not find any university acknowledgment of Koren's inappropriate professional behavior.

Yet Dr. Olivieri is the one whose name and professional standing were upended; fortunately, her story has been favorably reexamined, and she has become a hero and symbol of resistance to Big Pharma attacks.[29] The *Journal of Medical Ethics* (jmedethics) eventually applauded "Dr. Olivieri and a few of her research colleagues then at the HSC . . . [who] risked their careers, their health, and their finances to defend principles they held dear."[30]

In contrast, Charlie never received the science community's recognition of innocence and was locked out of his Northwestern office, basically "encouraged" to look for work elsewhere.

* * *

Another fantastic-but-true story of Big Pharma's reach into pharmaceutical retribution is the 1987 case of Betty Dong, drug researcher at the University of California, San Francisco. She investigated levothyroxine, sold as Synthroid. Its manufacturer Boots (British drug company) paid her to contrast their drug with similar ones on the market, hoping to prove theirs was superior. It wasn't. Although Boots had handpicked Dr. Dong, had specified the study design, and had made frequent quality assurance visits, executives suddenly objected to nearly all aspects of the study and complained to university officials. Boots even accused her of breach of contract, conflict-of-interest and ethical violations.[31] Two university investigations, however, found only minor and easily correctable problems.

Dr. Dong discovered that Boots had a contract with her university that stated she needed written permission to publish her conclusions. She also learned that an investigator had approached her university to learn if she had other deposits into her university funds.[32] She was amazed to read, four months later, a sixteen-page attack on her, her scholarship, and the conclusions of her unpublished article. According to researcher and investigator Stephen Fried, the author of that attack was Dr. Gilbert Mayor, who happened to be both the editor of that journal and also, coincidentally, the drug company's medical services director.[33]

Dr. Dong emphasizes the difference between her scientific colleagues' support and the administration's stance: "The University-wide faculty and my Department of Clinical Pharmacy supported my research. Support did not come from the legal arm of the university or the Chancellor's office. However, after the acquisition of Flint by Boots pharmaceutic, likely due to negative press and public perception, negotiation between the Chancellor's office and Boots allowed publication to proceed although 10 years later." When she was released from her university's nondisclosure agreement, she took her facts to the *Wall Street Journal*,[34] which then alerted the public to the dangers of Synthroid.

She finally published her scientific article[35] after the company relented "in the face of negative publicity and pressure from the Food and Drug Administration for possibly misleading claims."[36]

When Charlie published his findings on the bestselling drug for EPO, he was quickly the victim of a spurious series of attacks that questioned his honesty and professional integrity.

Surely the attacks were not coincidental.

* * *

Another widely publicized abuse of medical researchers was the attack on Dr. John Buse, University of North Carolina, who made several public presentations in 1999 about a new diabetes drug, Rosiglitazone.[37] The consequences he endured wound up on the floor of the US Senate Finance Committee in 2007.[38] In this case, the drug company, GSK, actually made direct contact with Dr. Buse's department head to keep him from publicly disclosing his findings. Buse felt his university forced him to sign various legal documents meant to prevent him from additional public disclosures. GSK had threatened him as personally liable for $1.4 billion. Only by signing the documents that required him to remain silent was he able to keep his job and avoid personal litigation. When the Senate Committee was alerted to Buse's predicament, they referred to Dr. Buse's treatment as "intimidation." Dr. Buse's major regret is that he wishes he had originally stood up to the company and perhaps saved more lives.[39]

* * *

Although they haven't received as much publicity in the United States, we uncovered stories of drug companies that threatened legal action over Italian public health officials' warnings about drugs;[40] they sued Spain's scientific bulletins for "committing scientific fraud" (but lost);[41]

an Australian university had to publish public apologies for believing its lecturer insulted a university/drug company product (Gardasil)—and for publicly rebuking the senior lecturer for a radio interview.[42]

Big Pharma—from Amgen to Johnson & Johnson—have lost much of the public trust they once enjoyed.[43] Stories of drug companies' willingness to ignore science, to abuse academic researchers, and to hide evidence proliferate; drug companies continue to increase essential drug prices.[44] Katherine Eban spent years interviewing over 240 participants in the generic drug production/marketing and reported a "labyrinthic story of how the world's greatest public health innovation [generic drugs] also became one of its greatest swindles."[45] Scandals continue; profits soar. It would be unwise to assume that all these abuses will soon be overcome by legislation.

It is not surprising that the creation of drugs seems to be shaped by different, more rigorous scientific standards than their marketing does. The former is controlled by scientists, the latter by business executives—two groups whose values and standards for success vary widely.[46] Where drug companies are involved, protectiveness can also be attributed to the huge monetary gains, promotions, and lifestyle benefits—all of which can be threatened by research scientists like Charles Bennett.

Universities: The Importance of Academic Freedom and Due Process

Normally, the university administrator's official tasks and ambition are to help the faculty member achieve the purposes of his grant. That's not what happened to Charlie. The decision by Dean Vaughn to lock Charlie out of his own professional office would be academically unacceptable *unless* the decisions were based on faculty review and its carefully arrived-at conclusion of serious misconduct—a conclusion that was notably absent. Charlie Bennett deserved an impartial investigation and a strong presumption of innocence. He received neither.

Had a case like Charlie Bennett's arisen during my faculty time at Indiana University, Stanford, or Yale,[47] I have no doubt that the decision to lock a scientist-professor out of his office would have been evaluated by a faculty committee, and there would have been vigorous faculty protest in response to public statements assuming guilt of a faculty member without due process. I was an active member of the American Association of University Professors (AAUP) at each of these institutions, and I was well aware (through public and private conversation) that the presidents of each of these distinguished universities respected the AAUP tenets of academic freedom. The overwhelming majority of faculty considered these concepts valuable and important.

At many levels, from account clerks to the university president, Northwestern employees and officials acted as though Charlie were guilty of significant misconduct. It took years to clear him of involvement in his administrator's ATSDATA fraud. What accounts for the disproportionate response? No one connected to Amgen is among the list of people who officially accused Charles Bennett, or made him an academic outcast, or locked him out of his office at Northwestern. Yet, without the secret involvement of an outside force, the entire deplorable episode is inexplicable. The coincidental timing of the attacks is the stuff of mystery novels. Charles Bennett had played a central role in undercutting EPO, and almost immediately he himself was undercut by his university.

Even as Dr. Charlie Bennett at Northwestern School of Medicine was severely punished for minor violations of his academic responsibilities, Dr. David C. Dale of the University of Washington School of Medicine was treated with respect despite acknowledging a ghostwritten "collaboration" only after he was accused and investigated. It is sadly illuminating to contrast the outcomes of the case against Dr. David Dale with that of Dr. Charles Bennett, who was essentially forced to leave Northwestern. At every step of the case, Northwestern officials chose not to follow the dictates of academic freedom to investigate senior faculty and took steps that increased his vulnerability: they cut him out of the settlement and refused to pay, or even help with, his legal bills during the crucial time period that followed.

Dale's case is not unique—academic institutions are often willing to wink and nod to stay on good terms with large donors. There are few, if any, examples of severe penalties for a practice that corrupts scholarship and endangers patients. The lack of genuine concern demonstrated by academic institutions has an easy, but unsatisfactory, explanation: a great source of income while a valuable partner in legitimate scholarship. Staying on good terms has benefits; battles will inevitably be costly.

The Justice System: The Importance of the False Claims Act

The False Claims Act played many roles in the cases discussed in this book. It was the legal instrument used by Melissa Theis, Linda Wyetzner, and Kurt Lindland to obtain millions of dollars from Northwestern University and half a million from Charlie Bennett. And yet. And yet it was also the vehicle by which Prof. Horwitz and the Attorney General's Office brought suit against Amgen, charging it with off-label marketing and ghostwriting of scholarly articles.

That suit led to Amgen's paying $760 million to settle the case. Plus, the Act was also the basis of Beverly Brown's action against Celgene, which was settled by Celgene's payment of over $300 million on the grounds of off-label marketing. Given this variety of claims and claimants, it is apparent that the Act can be both a deterrent to justice and also a weapon to fight injustice.

The case against Charlie Bennett involved an enormous amount of investigation by agents and agencies around the country and analysis by a team of elite accountants. With the exception of Feyi's ATSDATA fraud, which was quickly revealed, this massive investigation produced nothing that resembled an inappropriate grant claim for reimbursement of government funds. It may have been the case that some of

Charlie's consultants should not have been hired if he were carefully following government accounting guidelines. But failure to follow recommended procedures does not constitute a false claim. Generally, the federal agency goes after big cases of big crimes that offer real evidence. What happened throughout Charlie's ordeal was different: the AG office posted headlines, made endless threats and bullying over what was basically chump change from accounting errors. Government employees are, after all, on the taxpayers' tab. During this time, someone obviously had to be ignoring bigger crimes.

Despite Feyi's theft, the US Attorney's Office continued a lengthy investigation that revealed no action by Charlie that came close to meeting the definition of a false claim. Confusingly, there was no charge of financial-abuse conduct in the actual complaint (just items like "no prior approval of consultant"). If the United States does indeed have a "presumption of innocence," where was it in this case?

After the 2013 settlement that was carefully crafted to convince readers that Charlie was guilty of misusing government funds, Acting US Attorney Gary Shapiro issued a public statement warning researchers that they will be carefully audited.

> Allowing researchers to use federal grant money to pay for personal travel, hotels, and meals, and to hire unqualified friends and relatives as "consultants" violates the public's trust. This settlement, combined with the willingness of insiders to report fraud, should help deter such misconduct, but when it doesn't, federal grant recipients who allow the system to be manipulated should know that we will aggressively pursue all available legal remedies.[48]

In our later phone interview, Shapiro told me that he was not deeply involved, and *he knew of no actual incident* to support the damning accusations.

Amendments to the False Claims Act (FCA) in 1986 were specifically designed to encourage lower-level and midrange staff to report on misconduct by their superiors. It increased the amounts recoverable, and it protected whistleblowers from retaliation. The FCA is a profit center for the government, but here they were diddling with pennies and nickels. This wasn't a strongly enforced area like health care, where the government spends huge sums of money. Although considerable money can be made in these lawsuits (like Theis's payout), it's a drop in the bucket compared to lawsuits against pharmacies, for instance. On the other hand, there has developed a significant "false claims bar"—lawyers who make most of their living from accusations of false claims.

The case against Charlie interrupted his scientific career and brought him to the brink of bankruptcy.[49] Are Charlie's travails the result of weaknesses in the structure of the FCA? Or do they merely demonstrate anew that any statute that functions by imposing penalties on wrong doers is bound to produce mistakes and injustice from time to time?

The FCA, the Whistleblower Act, can be easily exploited. All it takes to invoke the FCA is a person with some claim to "inside knowledge" who would be willing to act as the whistleblower. It costs no money for a whistleblower to employ a contingent-fee attorney.

Whistleblowers tend to be treated as heroes rather than snitches by the public. When a large sum of government money is involved, the FCA actually provides treble damages.[50] As Charlie's case demonstrates, it is possible to put together a legally elegant complaint with no proof of wrongdoing—based on *possible* violation of government rules and failure to follow accounting standards with respect to expenditures. The price of a possible conviction may well force the defendant, even if innocent, to settle in hopes of returning to his or her former life. This happened to Charlie Bennett. Even being identified as a defendant in such a case will seriously harm a person's academic reputation and standing.

The FCA does require the claimant to show that the defendant sought to obtain government funding based on fraudulent misrepresentations that it "knew to be false." However, "knew to be false" turns out to be far too easy to distort.

The False Claims Act is in a state of flux and is expanding its reach about the concept of "knowingly false." In 2016, *In Universal Health Services, Inc. v. United States ex rel. Escobar*, the US Supreme Court unanimously held that liability under the FCA can be found based on the "Implied False Certification Theory."[51] In that case, a hospital did not inform patients that it was not in compliance with regulatory requirements. As a result of not having the proper medical personnel, the teenage daughter of the Escobar family died from a fatal seizure after experiencing adverse effects of a prescribed medication. The Supreme Court decided that the hospital was liable for the death under this "implied certification"—that is, that hospital officials gave the impression that they were in compliance with their rules of certification when they were not.[52] The hospital/defendant in that case had sought payment for services provided by unqualified personnel—despite an obviously serious, admitted failure to provide adequate service.

This case and its Supreme Court opinion have raised the issue of when failure to comply with government rules and regulations may be deemed "knowingly false."

The limits of FCA have not been established and, like many other criminal laws, it is in danger of being stretched beyond its original intention; if the "implied certification" doctrine were expanded to cover the failure to undertake independent, time-consuming research of these obscure, complex government regulations, a high percentage of university professors and scientists working under government grants would be vulnerable to false-claim allegations. As Charlie Bennett's case demonstrates, suspicion alone is likely to lead to a settlement that makes the relator and lawyer richer, and the defendant poorer.

Charlie's case makes clear that a thoughtful amendment of the FCA—to ensure that the "knowing falsity" is applied forcefully—would be valuable to protect scientists and academics from unfair attack.

CHAPTER 11

Charlie Today

As this book shows, Charlie Bennett has paid a price for his audacity of opinions and made powerful enemies, several of whom have made serious efforts to harm his career and to limit his effectiveness. He's been kicked off prestigious boards and commissions. His NIH grants have been suspended three times. He's been sued and forced to pay hundreds of thousands of dollars in damages for speaking out. He's been the target of a six-year Department of Justice investigation, which cost him $475,000 to settle although the investigation found nothing but accounting errors. Plus, he paid the Illinois agency in charge of medical licensing to not suspend him—even though there has not been a trial or admission of guilt. And he paid more than $600,000 in legal fees. That's a lot of punishment for someone whose case was dismissed or settled with no finding of guilt.

His struggles have taken a toll. He is no longer the easygoing optimist of his earlier years; instead, he anticipates trouble and worries about any future successes. But he has not given up his lifelong struggle for meaningful scientific achievement. Charlie, with his team, is speaking out publicly, once again. (As one impetus: Back in 2008, researcher Allison Grandey reported that European regulators were requesting that doctors should return to transfusion rather than ESAs. The article referenced Amgen and the Henke study—and recommended informed assessments of the benefits/risks with cancer patients.[1] Grandey is also stubborn in her attempts to warn of ESAs. She continues to warn the public about the cancer link.)[2] Today, Charlie has expanded those concerns to warning that Cipro and Levaquin antibiotics are dangerous and associated with bizarre suicides.

The connection between fluoroquinolones and suicides is shaky, primarily because other background factors can enter into a suicide. As of yet, no meta-analysis. Charlie is on the case yet again. As you read above, Charlie unsuccessfully petitioned the FDA to add suicide to the Black Box warnings for popular antibiotics: In 2022, he was interviewed about a Florida suicide that her family related to fluoroquinolones.

> Charles Bennett, medical expert, continues to petition FDA for checks on the popular antibiotic to protect public from suicide risk. He suspects that only 1–10% of all adverse drug side effects are ever reported.
>
> A nationally recognized medication safety expert, Dr. Charles Bennett, petitioned the FDA in 2014 and again in 2019 for increased suicide warnings for Cipro and Levaquin. They're in a class of drugs known as fluoroquinolones.
>
> In 2018 the FDA issued its strongest "black box" warnings for these drugs. It states that the antibiotics are "associated with

> disabling and potentially permanent side effects of the tendons, muscles, joints, nerves and central nervous system."
>
> *The agency denied Dr. Bennett's petition* to add suicide as part of the boxed warning, leaving it in a subsection on page 12 of the 52-page drug insert included with the prescription. [emphasis added to TV text.][3]

The fluoroquinolone patients' reporting doctors and families insist that this rash of suicides is committed by healthy women: several kissed their children goodbye and then used a gun,[4] or tied themselves to bricks and rolled into their backyard pool, or crashed their car into a brick wall, or walked deliberately in front of a semitrailer. Not normal.

Charlie has remained preternaturally active, applying for grants for drug safety problems that he believes the FDA should recognize. To date, he has been awarded more than $44.2 million in federal research grants and has started a pilot on cancer and COVID; a pilot project to evaluate COVID and pulmonary care; and a larger pilot project evaluating COVID-19 and respiratory care. He is working with Dr. Rosen on a pilot program to investigate the side effects of COVID, using an innovative approach—examining Twitter accounts instead of interviewing general practitioners to evaluate social media as a data source of safety signals with pharmaceuticals.[5] "That's what makes him special," Rosen told us. Charlie's 2022 idea has garnered publicity and praise and could lead other science/medicine researchers to follow suit.

Charlie's international SONAR group continues its twenty-five years of weekly meetings. We have listened in—fascinating stuff! A researcher in Spain shares a new case involving ESAs. Someone else has pulled the numbers on a bilateral survey. Charlie is in his element, for sure. For example, SONAR is now the leading group of investigators about a novel cause of acute leukemia, the concomitant use of an Amgen supportive care drug called Neupoen and Neulast. And the group continues to evaluate

Cipro-related neuropsychiatric toxicity. (Neuropsychiatric can be brain fog weakness, pain, etc.)

He continues to publish. His 2019 op-ed in the *Los Angeles Times*, "How the government failed us on opioids," draws a bright line of concern, that "the federal bureaucracy is simply not up to the task of adequately monitoring and controlling dangerous pharmaceuticals."[6] He faults the entire system.

Plus, he continues to share his findings and concerns with the medical field as well. His current research now details a multibillion-dollar drug, Rituxian, where SONAR has ironically found that safety concerns were overblown. The paper about this discovery, published in *Lancet Haematology*,[7] should lead to the removal of an unnecessary Black Box warning. He summarized the history of EPO and its new limitations with anemia.[8] He should know, right?

Charlie was there at the beginning of the "EPO frenzy" and helped end it. He has come to see himself as a leading representative of medical resistance to powerful and wealthy drug companies and their prestigious medical bureaucrats.

In a 2022 study that gives him great satisfaction, he has gathered major research scientists together to expose pharmaceutical companies that have attacked researchers—an article both astonishing and terrifying. In "David versus Goliaths: Pharma and Academic Threats to Individual Scientists and Clinicians,"[9] he and twenty authors reported on twenty-six individual researchers who felt threatened or intimidated by corporate officials. (Several of these authors were mentioned in detail above, chapter 10.) The article names names and describes documented punishment for those researchers who went against "the message": threats, intimidation, harms including loss of employment, job demotion, delayed academic tenure decisions, and personal payment of legal fees for individual clinicians/scientists after public findings.

Today, Dr. Charlie Bennett still bears scars from his battles. Every aspect of his professional life has been blistered and scraped and nicked, and that goes double for his good name and personal reputation. He hopes that somehow, someone will tell his story accurately, fairly, and truthfully.

We have followed his triumphs and failures and persistence. When in 2021 we called to inform him that Skyhorse Publishing intended to publish this book, we found him aboard an Antarctic cruise, file folder in hand, regaling passengers and crew with his story and declarations of innocence.

Notes

Chapter 1

1. Dr. Stephen Rosen. Telephone interview with author, April 3, 2017.
2. The grants included a Veterans Affairs Career Development Award to study the Quality of Care for Patients, a Research and Development Merit Review Award of $525,000 from National Institutes of Health, and $780,000 for a four-city study from the National Center for Health Services Research.
3. Some years later, Jan Meister appeared on the PBS program *Chicago Tonight* in May 2009 and described how Dr. Charlie Bennett's intervention had saved her life. She also donated money to Northwestern School of Medicine to help support Charlie's developing research, the Jan Meister Grant for Drug Safety at Northwestern University School of Medicine, in July 2009.
4. Charles Bennett et al., "TTP Associated with Ticlopidino: a Review of 80 Cases," *Annals Int. Med.*, April 1990; Charles L. Bennett, MD, PhD; Charles J. Davidson, MD; Dennis W. Raisch, PhD; et al. Original Investigation: "Thrombotic Thrombocytopenic Purpura Associated with Ticlopidine in the Setting of Coronary Artery Stents and Stroke Prevention," November 22, 1999, https://jamanetwork.com/journals/jamainternalmedicine/fullarticle/485186.

5. Tom Burton, "Hematologist Says Data Implicate Ticlid on Cases of Blood Disorder," *Wall Street Journal* (March 31, 1998). www.wsj.com/SB891310432860249500.
6. Among his grants was one of $2,500,000 from the National Institutes of Health/National Heart, Lung, and Blood Institute, R01 to study "Case-Control Study of Ticlopidine-Associated Thrombotic Thrombocytopenic Purpura." 9/9/02–8/31/07 *NCE*. HL069717.
7. Bennett et al., "Association Between Pharmaceutical Support and Basic Science Research on Erythropoiesis-Stimulating Agents," pubmed.nlm.nih.gov/20837837.
8. FDA added Plavix warning (March 12, 2010). www.fda.gov/drugs/postmarket-drug-safety-information-patients-and-providers/fda-drug-safety-communication-reduced-effectiveness-plavix-clopidogrel-patients-who-are-poor. *Also see*: "Brand Drug Preemption Shuts Down Plavix Litigation in New York State Court" (Jan 11, 2018), www.druganddevicelawblog.com/2018/08/brand-drug-preemption-shuts-down-plavix-litigation-in-new-york-state-court.html. Plavix is a drug prescribed to inhibit the formation of blood clots. As such, ever since it has been on the market, its label has included warnings regarding the risk of bleeding. *In re: Plavix Products Liability Litigation*, 2018 WL 4005859, (N.Y. Sup. Aug. 22, 2018). It is that same risk which plaintiffs in the litigation allege was insufficient. Defendants moved for summary judgment arguing plaintiffs' failure to warn claims were preempted because defendants could not independently have changed the Plavix warning and plaintiffs' design defect claims were preempted because defendants could not have changed the design (i.e., the chemical composition) of an FDA approved drug.

Chapter 2

1. The Cochran collaboration involves scientists working together to address key health questions, and it is also an effort to collect scientific information such as reports of experiments and make them available. It is based on the ideas of Archibald Leman Cochrane CBE (12 January 1909–18 June 1988).
2. Failure of the kidney to produce enough EPO causes anemia, which requires regular blood transfusions. This was particularly worrisome in the 1980s when transfusions were understood to create a severe risk of acquiring AIDS.
3. Recombinant technology involves creating a new substance by combining the DNA of various existing substances. It is genetic material formed by reconfiguring other genetic material. "Recombinant DNA, molecules of DNA from two different species that are inserted into a host organism to produce new genetic combinations that are of value to science, medicine, agriculture, and industry." https://www.britannica.com/science/recombinant-DNA-technology.

4. The relationship between Amgen and Ortho was from the start competitive and stormy. Ortho accused Amgen of seeking to lock it out of the market of anemic kidney disease patients; Amgen accused Ortho of marketing their brand of EPO (termed PROCRIT) to doctors who were providing care to anemic patients on dialysis. Both were correct. With lawsuits, arbitrations, the hiring and training of salespeople, and the development of new facilities, both companies made huge marketing and sales expenditures that required intense sales efforts to generate revenue. Edmund L. Andrews, "Amgen Wins Fight Over New Drug," *New York Times*, March 7, 1991.
5. Amgen's future was set when it won that patent lawsuit, recognizing Amgen's sole right to manufacture and sell recombinant EPO. And, boy, did it sell. After its victory, Amgen's stock soared from $12 per share to $113, a clear indication of the value of EPO to whoever held that patent. Goldwasser long lamented his own failure to apply for an individual patent, and he never profited significantly from his twenty-year role in the development of EPO. Goldwasser Family Trust, *A Bloody Long Journey* (privately published 2001) (from Eugene Goldwasser Self Declaration 7/28/86).
6. "Pure red cell aplasia (PRCA) is a rare disorder of blood production in which the bone marrow, the spongy tissue in the center of the bones, fails to function in an adequate manner resulting in anemia. Red blood cells are responsible for carrying oxygen to the entire body." https://my.clevelandclinic.org/health/diseases/14475-pure-red-cell-aplasia-prca. Visited June 2020. https://pubmed.ncbi.nlm.nih.gov/15792922/. Visited July 2020.
7. The problem appeared to be an immunologic problem that resulted with subcutaneous administration of the J&J product that contained polysorbate 80 at a certain concentration. Those countries that were reporting the negative statistics had been using the polysorbate-containing J&J product—and were administering it subcutaneously.
8. Bennett et al., "Association Between Pharmaceutical Support and Basic Science Research on Erythropoiesis-Stimulating Agents," *Arch Inter Med* 170 (16) (2010): 1490–8.
9. Roy Baynes email to 6 Amgen employees, Feb. 20, 2005. On file with authors.
10. Doctors who give speeches and write journal articles are Key Opinion Leaders (KOL). See full, fascinating discussion: Sergio Sismondo, *Ghost Medicine: Big Pharma's Indivisible Hands*, 14–18. He summarizes that KOLs are the are the "zombies of the industry" because "their brains and souls have been taken over." 16.
11. Kathleen Sharp, *Blood Medicine: Blowing the Whistle on the Deadliest Drugs Ever.* (Penguin, 2012). Sharp's entire book reveals step-by-step abuses of the pharmaceutical companies and their nefarious relationship to doctors who prescribe all these drugs.

12. Michael Henke, "Erythropoietin to treat head and neck cancer patients with anaemia undergoing radiotherapy: randomised, double-blind, placebo-controlled trial." *The Lancet* 362 (2003): 1255–1260.
13. Goldberg's *The Cancer Letter* is an award-winning cancer newsletter sold by subscription. He achieved a high level of accuracy by consultation with leading oncologists.
14. "Do Erythropoietin Receptors on Cancer Cells Explain Unexpected Clinical Findings?" 10:24 (29) *Am. Soc. Clinical Oncology,* Sept. 21, 2006:4708–13.
15. In *Lancet Oncology*, Aug. 4, 2003, (8): 459–60.
16. Brian Leyland-Jones, "Erythropoiesis Stimulating Agents: A Personal Journey," *JNCI: Journal of the National Cancer Institute*, 105, no. 14, July 17, 2013: 999–1001, https://doi.org/10.1093/jnci/djt171. And see https://academic.oup.com/inci/article/105/14/999/96533.
17. The study was a multicenter, double-blind, randomized, placebo-controlled study.
18. Paul Goldberg, "FDA Approves Aranesp for Chemo-Induced Anemia."; Paul Goldberg, *The Cancer Letter* 257 (8) Sept. 30, 2002, "Aranesp Improved Anemia in Phase II Trials." Paul Goldberg, *The Cancer Letter*, Feb. 15, 2007, Paul Goldberg, *The Cancer Letter* 28 (n9) August. 21, 2002, Paul Goldberg, *The Cancer Letter* 34 v. 36, Oct. 3, 2008, "Amgen Announces Cochran Findings: ESAs Increase Risk of On-Study Death, 1.
19. Charlie learned of the Dahanca 10 results from Goldberg. In an article on the Dahanca study that Goldberg published in *The Cancer Letter*, Goldberg accused Amgen of failing to disclose the results of the study to the public. https://cancerletter.com/free/20150904_2.The results were dramatic. As described by Goldberg: "After my story was published, Amgen's stock crashed, precipitating a shareholders lawsuit." Feb. 27, 2017. case-law.vlex.com/vid/goldberg-v-amgen-inc-893675714.

    In 2015, Goldberg refused to respond to a subpoena from Amgen asking for information on whether he had leaked his findings to Wall Street analysts, and his position was upheld in an excellent opinion by Judge Ahmit Mehta, holding that a reporter's First Amendment privilege applies to all conversations, not merely those undertaken with a pledge of secrecy. He concluded that Amgen did not take adequate steps to find the information on its own. Goldberg and his lawyers made clear that they saw Amgen's action in seeking to depose him as payback for his criticism. htts://rothwellfigg.com/results/us-district-grants-motion-in-important-First- Amendment-case, Aug. 21, 2015.
20. Cochrane Data Base Syst. Rev. 8 (3) (July 19, 2009). CD001431. Also published: Bohlius, J., Wilson, J., Seidenfeld, J., Piper, M., Schwarzer, G., Sandercock, J., Trelle, S., Weingart, O., Bayliss, S., Djulbegovic, B., Bennett, C. L., Langensiepen, S., Hyde, C., and Engert, A. "Recombinant human erythropoietins and cancer patients: updated meta-analysis of 57 studies

including 9353 patients." J Natl Cancer Inst. 2006 May 17;98(10):708–14. doi: 10.1093/jnci/djj189.

21. Charlie's analyzed data showed that EPOs increased mortality by 10 percent and blood clots by 60 percent and that these findings were independent of whether a target hemoglobin or 10 or 12 mg/dl was used. This new analysis reinforced Henke, Leyland-Jones, and Dahanca 10 but was also a significant addition to them because the earlier studies had used very high doses of EPO and DARB as they targeted hemoglobin levels of 12 mg/dl or higher.
22. These important results were not published in final form in a peer-reviewed medical journal until 2017.
23. *The Cancer Letter*, June 1, 2006, 1.
24. Baynes, MD, PhD, is currently Senior Vice President and Head Global Clinical Development, Chief Medical Officer, Merck Research Laboratories; in 2007 he was Vice President of Amgen of Clinical Oncology.
25. Amgen had earlier received a "confidential embargoed copy" of his paper. "Please note that these were provided by a reporter under confidentiality, so please do not forward or distribute," wrote Amgen's Trish Hawkins to a large group of senior employees. June 8, 2007. On file with authors.
26. Perhaps they were responding to Bill Sheridan's email of May 31, 2007, "Bennett meta-analysis/implications of comment letter/next actions." Sheridan encouraged his readers: "We need an answer to the question: Why is Bennett wrong and Amgen right? Why should we believe Amgen?"
27. Susan Kelly and Debra Sherman, "Anemia drug safety debate overblown: doctors," (June 3, 2007). reuters.com/article/us-cancer-epoetins-3-idUSN034 1927520070604.
28. Wright, J. R., Ung, Y. C., Julian, J. A., Pritchard, K. I., Whelan, T. J., Smith, C., et al., "Randomized, double-blind, placebo- controlled trial of erythropoietin in non-small-cell lung cancer with disease-related anemia, *J Clin Oncol.* 24 (29), 2006 Oct. 10:4708–13. ; *J Clin Oncol* Epub 20 March 2007 [PubMed].
29. Bennett, C., Henke, M., Djulbegovic, B., et al., "Venous thromboembolism and mortality associated with recombinant erythropoietin and darbepoetin administration for the treatment of cancer-associated anemia." 299 *JAMA,* Dec. 4, 2008: 914–924.
30. Andrew Pollack, "F.D.A. Panel Seeks Limits on Cancer Patient Drugs," *New York Times* (May 11, 2007): https://www.nytimes.com/2007/05/09/business /09anemia.html.
31. Otis Webb Brawley, with Paul Goldberg, *How We Do Harm: A Doctor Breaks Ranks About Being Sick in America*, St Martin's Press, 2012.
32. Brawley, 244.
33. See, for instance: J. Douglas Rizzo, Mark R. Somerfield, Karen L. Hagerty, Jerome Seidenfeld, Julia Bohlius, Charles L. Bennett, David F. Cella,

Benjamin Djulbegovic, Matthew J. Goode, Ann A. Jakubowski, Mark U. Rarick, David H. Regan, Alan E. Lichtin, "Use of Epoetin and Darbepoetin in Patients With Cancer: 2007 American Society of Clinical Oncology/ American Society of Hematology Clinical Practice Guideline Update." (2007). Also the FDA decision was reported in *The Cancer Letter,* March 17, 2008. An important analysis.

34. Peter Whoriskey, "Anemia Drugs Made Billions but at What Cost?," *Washington Post,* June 19, 2012.

Chapter 3

1. Alice Comacho, 2004 college graduate, was hired in 2004. In 2006, Licht appointed her as Bennett's account clerk. I called Camacho, who made clear she did not want to discuss her office interactions with Dr. Bennett. Alice Comacho, brief interview with author, 2008.
2. It is true that Comacho had some basis for concern: at one point, she was besieged by Bennett's mentees seeking payments and discovered that RADAR had no available funds. On Nov. 8, for instance, she denied a RADAR request to attend a conference. Department of Health and Human Services (DHHS). 12/14/09. On file with author.
3. Camacho, Department of Health and Human Services (DHHS). 12/14/09. On file with author.
4. Angela Youngfountain was Comancho's supervising purchasing coordinator.
5. Dr. Knight credits Charlie's mentorship with altering her path from clinician to scholar.
6. Attorney Peter Winn, interviewed by author, May 10, 2018.
7. With Bart Stupak, chair of the Oversight Committee, Dingell also investigated generic drug scandals: In *Bottle of Lies: The Inside Story of the Generic Drug Boom,* (HarperCollins 2019), Katherine Eban meticulously traces corrupt FDA employees through private investigators to sacks of trash and finally onto the desk of Rep. John Dingell (D-MI). Eban: 222–225. "Dingell's committee uncovered corruption that seemed to have no bottom. Generic drug executives had roamed the FDA/s halls, dropping envelopes stuffed with thousands in cash onto the desks of generic drug reviewers. . . . A generic drug trade association had subsidized the appearance of FDA reviewers at conferences. . . . Rep. Ron Wyden (D-OR) called the generic drug industry 'a swamp that must be drained.'" Sadly, the murky waters remain.

Chapter 4

1. United States of America ex rel. Melissa Theis and Melissa Theis, individually v. Northwestern University, The Robert H. Lurie Comprehensive Cancer Center of Northwestern University, Dr. Steven T. Rosen, and Dr. Charles L.

Bennett, United States District Court, Northern District of Illinois, Eastern District (May 2009) under seal.

2. Amgen later pled guilty and settled the case in 2012 for $760 million. That settlement included a criminal finding against Amgen with a $160 million fine and a civil settlement of $600 million—the largest case involving a biotechnology company and the Department of Justice, ever. "Amgen Inc. Pleads Guilty to Federal Charge in Brooklyn, NY.; Pays $762 Million to Resolve Criminal Liability and False Claims Act Allegations | OPA | Department of Justice," justice.gov/opa/pr/amgen-inc-pleads-guilty.
3. Rosenbloom refused to discuss the case with me, years after it closed.
4. Charlie suspects that Dennis West, motivated by the ambition to succeed him as the head of RADAR, informed the department's officials that Charlie had thrown papers into the trash. Charlie believes that West reported that he violated the rule in order to force Charlie out and take over as head of RADAR, which he did.
5. I have participated in many academic cases, much weaker than Charlie's, in which the defendant was able to present facts and effectively challenge a proposed penalty.
6. Boris Pasche, telephone interview with author, October 10, 2018.
7. Fred Goodwin, telephone call to Charlie Bennett, 2009.
8. Fred Goodwin, telephone call to Charlie Bennett, 2009.

Chapter 5

1. His former mentees include Sara Knight, PhD, now a professor of medicine at the University of Utah School of Medicine. She credits Dr. Bennett with playing a key role in achieving her desired career as a scholar. She told us, "I would never have had the career I yearned for without Charlie Bennett."
   Dr. June McKoy of the Feinberg School of Medicine is similarly grateful. "He not only mentored me; he sponsored me." Another budding scholar supported by Bennett's grants was Elizabeth Calhoun, who worked with Bennett's team for several years inside the VA hospital's ICU units. For several years she was PhD, MEd, executive director of the Center for Population Science and Discovery at University of Arizona Health Sciences, with a joint appointment at the University of Arizona College of Public Health. She credits Bennett with making a major contribution to her academic success. Plus this:
   "He was a wonderful mentor," David Friedland, now a professor at Harvard and chief medical officer at Blue Cross/Blue Shield of Massachusetts, told us.
2. [Anonymous] telephone call to Charlie Bennett, 2011.
3. Amy Bennett interviewed by author, Charlotte, SC, 2018.
4. Franz Kafka, *The Trial* (Schokea Books 1925), 2.

Chapter 6

1. Melissa Theis, notes to Power Point Committee.
2. Quishun Elrod, FBI summary interview, 7/31/09 on file with author.
3. Otis Brawley, telephone interview with author, 2012.
4. Professor Elizabeth Calhoun, interview with author, Palm Spring, CA, 2021.
5. Steve Rosen, interview with author, Palm Spring, CA, 2021.
6. hcrenewal.blogspot.com. Visited March 2020.

   In contrast with Poses's estimate is the brief statement made by David Stetler in our brief preliminary, never-followed-up, phone conversation: "I understand that he feels like he was treated wrongfully, but I think if you talk to anybody who's familiar with the facts, you would come to the opposite conclusion." David Statler, telephone interview with author, Jan. 11, 2017.
7. Roy Poses, "The Mystery of the Northwestern Settlement," *Healthcare Renewal,* Aug. 5, 2013. blog: herenewal.blogspot.com/search?q=2013+Northwestern. Also on file with author.
8. United States of America v. Feyifunmi Sangoleye. Case Number: 13 CR 528–1. Case: 1:13-cr-00528 Document # 24 Filed: 03/28/14.
9. 2019 PhD in Philosophy, Nursing Science, and a 2012 MSN in nursing from DePaul University. She stayed close to home—remember, she had used the University of Chicago as the basis for her fraud scheme.
10. Kurk Lindland, telephone interview with author, 2017.
11. Gregory Pratt, "Doctor working for Cook County tallied $248K in unauthorized expenses—including piano, flights, iTunes charges," *Chicago Tribune,* April 18, 2018: chicagotribune.com/news/breaking/ct-met-cook- county-doctor-expense -reports-20180413-story.
12. Inevitably Charlie's friends and coworkers were confused and frightened by the process. Charlie remembers that Nick Bandenko wrote him an email, and the email was really frightening. He goes, "Charlie. There are FBI agents at my door now. Call me right away." I went over there and said, "Whatever they're at your door for, tell them the answers to anything they ask." He was scared—who gets a knock from the FBI? He said, "I can't get involved in that kind of a thing, you know." Email to Charlie Bennett, 9/16/2010. On file with author.
13. Dennis Cournoyer, HHS file summary, on file with author, Sept. 16, 2010.
14. Johnathan Licht, HHS file summary, on file with author, Sept. 5, 2012: 2,3.
15. AUSA Kurt Lindland, telephone interview with author, 2017.

Chapter 7

1. She knew about the event because her sister Karen Weiss, MD, who at that time worked for the FDA, had been appointed Director of the FDA'S Safe Use Initiative.

2. John Fratti, telephone interviews with author and taped story, on file with author, February 27, 2019.
3. Sidney Wolfe, MD, cofounder with Ralph Nader of Public Citizen's Health Research Group, filed a lawsuit against the two companies in 2006, claiming that the two drugs were responsible for hundreds of persons developing Achilles tendon ruptures. The suite was not widely publicized. The case was initially dismissed, but Dr. Wolfe continued his effort to change the labeling of the drugs. In 1990, he published *Worst Pills Best Pills* (Pocket Books/Simon & Schuster). Wolfe was a determined health advocate: previously, it had taken him thirty years to get Darvon off the market; his involvement against the two pharmaceuticals helped this case achieve an earlier outcome. Stephen Fried offers a balanced and thorough review of Wolfe's work and reactions by his fans ("enormous resource") and detractors ("loose cannon") in *Bitter Pills: Inside the Hazardous World of Legal Drugs* (Bantam 1998), 178–180. Also see: Max Fitzer, *The Big Pharma Conspiracy: the drugging of America for fast profits*: (Make Profits Easy LLC, 2015). The FDA Transparency Initiative went so far as to bar Wolfe from a panel because he had advised readers not to take Yaz (Bayer's birth control pill) because of limited data. Instead, they allowed four other panelists with ties to Bayer. 94.
4. David Melvin. Telephone interview with author, 2019. The two earliest websites for fluoroquinolone victims were Melvin's and Linda Martin's flouroquinolonestories.com and myquinstory.info.
5. See Marcia Angell, *The Truth About Drug Companies: How They Deceive Us and What to Do About It* (Random House 2005), 150+, describing victim patient groups and problems with their being infiltrated by company "supporters" who change lobbying language, etc.
6. Laura, flouroquinolonestories.com, visited July 2020.
7. http://www.fluoroquinolonestories.com/2017/09/25/marianne-cs-story / Visited August 2020. One of 147 published web stories reveals that Cipro's adverse reactions continues in 2020:

   My name is Marianne and on April 11, 2017, I was given 500 mg of Ciprofloxin for a sinus infection . . . .

   As of today, September 14, 5 months since I was floxed, I am still suffering with extreme fatigue, and peripheral neuropathy in both hands, arms, feet, legs, face, and tongue. Muscle weakness, Muscle pain in my thighs, arms. My joints are stiff and painful. I have a problem with sleeping, heart palpitations, periodic muscle twitching all over, muscle tremors & jerks, lack of appetite, weight loss, headaches, balance issues. I am getting really tired of this . . . .

   I continue on this journey and prayerfully wait for improvements and/or complete healing.
8. After the initial FDA action in 2013, tragedy struck the University of South Carolina science faculty and the first work of the fluoroquinolone study. In

February 2015, one week after completing the manuscript describing the laboratory findings of liver toxicity with Cipro, Raja Fayad was shot and killed by his estranged, jealous wife.

9. Linda Martin. Telephone interview with author, Feb. 26, 2018.
10. Subsequent meeting among Bennett, author, and Martin, 2018.
11. The FDA probably knew how Charlie and RADAR felt about its process. In 2007 they had published "Evaluation of serious adverse drug reactions: a proactive pharmacovigilance program (RADAR) vs safety activities conducted by the Food and Drug Administration and pharmaceutical manufacturers," PubMed (nih.gov) ncbi.nim.nih.gov/17533207/.
12. Golomb, B. A., Koslik, H. J., Reed, A. J. "Fluoroquinolone-inducing serious, persistent, multisymptom adverse effects." *BMJ* Case Rep. 2015 Oct. 05 PMID: 26438672.
13. Linda Martin, telephone interview with author, 2019. (Also see chapter Charlie Today, above.
14. Dr. Golomb is notable for her theory of mitochondrial toxicity, which helps explain the negative reactions that so many patients had to fluoroquinolones. Beatrice Golomb, "Statin Adverse Effects: A Review of the Literature and Evidence for a Mitochondrial Mechanism," *Am J Cardiovasc Drugs* 8, no. 6, November 1, 2008: 373–418.
15. The FDA, the Department of Health and Human Services, Public Health Service, Food and Drug Administration, and Center for Drug Evaluation and Research "FDA Drug Safety Communication: FDA requires label changes to warn of risk for possibly permanent nerve damage from antibacterial fluoroquinolone drugs taken by mouth or by injection," April 15, 2013, Fda.gov /media/86575.
16. FDA's strongest warning. Doctors and patients need to be better alerted to these warnings; in despair, Stephen Fried titled one chapter of *Bitter Pills* "Nobody Reads the Label Anyway": 80.
17. Charlie had chosen this citizen petition route because in 2005, then Connecticut Attorney General Richard Blumenthal and Charlie had filed the first ever successful citizen petition to the FDA from a collaboration of a state attorney general and a medical researcher.
18. The warning: "Possible Mitochondrial Toxicity Fluoroquinolones, including Levaquin, may cause Mitochondrial Toxicity due, in part, to an insufficiency of ATP. Mitochondrial conditions that are due to an insufficiency of ATP include developmental disorders of the brain, optic neuropathy, neuropathic pain, hearing loss, muscle weakness, cardiomyopathy, and lactic acidosis. Neurodegenerative diseases like Parkinson's, Alzheimer's and amyotrophic lateral sclerosis (ALS) have been associated with the loss of neurons due to oxidative stress generated by reactive oxygen species (ROS) related to Mitochondrial Toxicity. Peripheral neuropathy, hepatoxicity, glucose disturbances, and phototoxicity may result from Mitochondrial Toxicity."

19. "Mitochondrial Toxicity Warning Urged in Petition." www.aboutlawsuits.com/levaquin-mitochondrial-toxicity- warning-petition-68138.
20. R. Ryne Danielson, "Watchdog Group Advocates for Sterne warnings on Antibiotics Levaquine, Cipro," *The Catalyst* 33 #31, April 2015, 1. *The Catalyst* is the medical newspaper of the University of South Carolina.
21. The FDA met with academic advisers; FDA staff; senior scientists from J&J and Bayer; persons who had been injured by Cipro or Levaquin; Sidney Wolfe (who led Ralph Nader's Public Citizen Health Research Group); attorneys for persons who were purportedly injured by Levaquin or Cipro; and Charlie.
22. "Another petition filed earlier in the year in collaboration with the Southern Network on Adverse Reactions (SONAR) requested a black box warning for mitochondrial toxicity." https://topclassactions.com/lawsuit-settlements/lawsuit-news/44582-fda-citizen-petition-wants-psychiatric-risks-added-quinolones-label/ Visited March 2020.
23. Charlie sought a process that could predict psychiatric syndromes association with the drugs. Bennett, Martin, and another scientist at the University of South Carolina filed a provisional patent application based on whole genome sequencing that identified a specific abnormality among 13 of 24 persons with severe neuropsychiatric toxicities. Bennett A, *Federal Practitioner* 2017.
24. Linda Martin, telephone interview with author, 2019.
25. ABC Channel 2 Action News (WSB-TV) show, Cleveland news 7/22/15. "Leviquin." Youtube.com/watch?. E3LILHpgrbc and cipro. PoisonDotCom. (featured Charlie and the active Flox group).
26. Ron Regan, "Popular antibiotic linked to deaths," CBS News, March 2, 2015, (before hearing); Ron Regan, "Levaquin: FDA fails to disclose additional serious side effects of antibiotic linked to deaths," CBS News, (July 27, 2015; Ron Regan, Sept. 23, 2016: www.news5cleveland.cm/local-news/investigations/family-of-mentor-woman-suspect-popular-antibiotic-prompted-suicide-death.
27. Linda Martin, telephone interview with author 2019. On file with author. 2019. Cipro victims continue to reach out to Charlie. In 2017, a Public Health Sciences educator from Clemson appealed to Bennett for any updates and referrals to a doctor who could help her brother. Paige Hooper. Email June 5, 2017, on file with author. In 2020, a mother's plea also referred to the Flox group and Charlie. Kara Kenney, "Indiana mother's fight for drug label change hits snag after FDA denies petition," WRTV, Sept. 2020.
28. Linda Martin, telephone interview with authors, 2019.
29. PONS. (English translation) "Antibiotic Stewardship," Dec. 8, 2015, https://link.springer.com/article/10.1007/s00120-015-0012-2. Visited March 2020.
30. "Disabling and potentially permanent side effects lead to suspension or restrictions of quinolone and fluoroquinolone antibiotics," Nov. 16, 2018.

ema.europa.eu/en/news/disabling-potentially-permanent-side-effects-lead-suspension-restrictions-quinolone- fluoroquinolone.

31. http://www.myquinstory.info/another-milestone-crossed-fluoroquinolones-the-fda-and-psychiatric-adverse- events posted July 14, 2018. Visited March 2020.; "FDA Denies Petition: Purdue student Shae McCarthy died 2013," www.wrtv.com/news/call-6-investigators. The student died after taking Levaquin; the FDA continues to deny Dr. Bennett's petition for label changes.

Chapter 8

1. Well after the case was settled, Lindland made clear to me that he continued to believe that Charlie was "in on it with respect to the ATSDATA fraud."—Kurt Lindland. Phone interview with author, Spring 2017.
2. Steve Rosen, initial interview with author, April 12, 2019.
3. August 9, 2013.
4. According to their web information, G&P provides "trial lawyers, their clients, receivers, business owners and others with professional services related to dispute resolution, commercial/business litigation, bankruptcy proceedings, family law proceedings and investigations." https://www.litcpa.com/company-profile. Visited June 2020.
5. "[Carla Pugh] directed the Clinical Simulation Program at the University of Wisconsin Hospital & Clinics, where she was also vice chair of education and patient safety. For her innovative work, she received the Presidential Early Career Award for Scientists and Engineers. Carla was also a member of TEDMED 2015's Editorial Advisory Board. She is now Professor of Surgery, at Stanford University School of Medicine. She is also the Director of the Technology Enabled Clinical Improvement (T.E.C.I.) Center." http://profiles.med.stanford.edu.
6. E.g., Arabyat, R. M., Raisch, D. W., McKoy, J. M., Bennett, C. L. "Fluoroquinolone-Associated Tendon-Rupture: A Summary of Reports in the Food and Drug Administration's Adverse Event Reporting System." *Expert Opinion on Drug Safety* 14 (11), November 2014: 1653–60. https://doi.org/10.1517/14740338.2015.1085968 .
7. This statement is still retrievable from the Northwestern University website. https://news.northwestern.edu/stories/2013/08/message-from-northwestern-university-president-morton-schapiro-provost-dan-linzer-and-feinberg-school-of-medicine-dean-eric-neilson/ (accessed 4/11/2021).
8. Howard Brody, "Hooked: Ethics, Medicine, and Pharma," brodyhooked.blogspot.com/2012/05/hiding-truth-about-drugs-how-long-are.html. "As [Steve] Goldberg recounts, Northwestern had previously had problem in this area, and a decade ago had to repay the NIH $5.75M that it obtained through inflated reimbursement for faculty effort. Just recently, Northwestern agreed

to pay the Feds a settlement of about $3M due to questionable payments on another grant, in this case one whose principal investigator was Dr. Bennett."

9. Northwestern University payment of $5.5 million to resolve False Claims Act and Common Law allegations (justice.gov). oig.nasa.gov/docs/pr2009-B.pdf. (see Appendix B).
10. Bob Kenney, NACUA Fall Workshop. "Legal Issues in Higher Education Sponsored Research, Compliance and Technology Transfer. 07B. Enforcement of Federal Grant Accounting: The Legal Perspective," November 15–17, 2006, xxv-06-11-8.doc (live.com).
11. *The Cancer Letter* (V.40, No.41), Oct. 31, 2014: 13.
12. All activity Dr. Bennett performed as PI were prior approved or approved at several different levels by Northwestern and the Robert Lurie Cancer Center. It is uncontested that Northwestern University and/or the Robert Lurie Cancer Center was the Grantee/Recipient on all of the grants that were the subject of the qui tam. By law/regulations, however, the Grantee (and not Dr. Bennett) assumes all legal responsibility and is accountable to the NIH for the grant's performance and is the sole legal entity responsible and accountable to the NIH for the performance. All financial aspects of the grant activity Dr. Bennett performed as PI were prior approved or approved at several different levels by Northwestern and the Robert Lurie Cancer Center. (see Appendix C: NIH Grants Policy Statement and Application Guide).
13. hcrenewal.blogspot.com. Visited March 2020.
14. Howard Kline. Telephone interview with author, 2018.
15. In contrast with Poses's estimate is the brief statement made by David Stetler in our brief preliminary, never-followed-up, phone conversation: "I understand that he feels like he was treated wrongfully, but I think if you talk to anybody who's familiar with the facts, you would come to the opposite conclusion." David Statler, telephone interview with author, Jan. 11, 2017.
16. Steve Rosen. Telephone interview with author, 2018. Rosen added a story of his own failure to conform to government regulations:

    "It happened several years ago. I was at the Veterans Hospital, and I had honestly accomplished more at the Veterans Hospital in a year than most people had in their entire careers. And there was a colleague whose wife was pregnant, and he asked me at the last moment, would I go to North Dakota in the winter to give a lecture on his behalf? And I did him a favor. I went there. I stayed at a Ramada Inn. It was minutes off the corridor someplace. And it turned out a pharmaceutical company was sponsoring the trip. I didn't know that at all. I just thought I'd do a North Dakota Medical Association, blah, blah, blah. I get a $500 honorarium for it. I come back. At the time in the VA, I was probably assigned to work 35 hours a week, and I was probably working 80 hours a week. They had these regulations we had to follow if we were going somewhere. I didn't even know about it because the document they give you is fifty pages long—who reads it? And I'm not thinking about that at all.

So, it turns out there are two physicians in Pennsylvania who had indeed abused the VA. They had taken much money outside their work, so they were investigating everybody at the VA. And they send around a letter saying, 'Did you take any time, blah, blah, blah, did you ever get an honorarium?' And I was one of the few people who answered honestly. Colleagues of mine told me afterwards, 'You responded to that? You have to be crazy!'

So the Head of Medicine responded. They brought me into a room with a new lawyer and an older lawyer. The older lawyer was like those who were involved with Charlie. He was clearly, right by the book. He tells me, 'You took a day off from the VA. You didn't sign out properly. You got an honorarium from, I guess, Eli Lilly. We have to give you a warning; what you did was wrong.'

"And anyway, I go through this whole routine with the new lawyer; she said to me, "Who are the schmucks who would put you through this, for everything you've done, and the amount of time you've put it? Do they recognize this is not something you're doing intentionally to harm the VA in any way or do something that's devious?' She recognized how absurd this was. And the complete stupidity of these people. But they can't distinguish between what is important versus what is not important."

17. McGurk tried to keep Amy abreast of the multiple parties and filing. (see Appendix D).
18. Illinois state law cannot protect against federal income tax bills, either. The Illinois Homestead Exemption: Breaking Down Five FAQs (gundersonfirm .com).
19. Charlie recalls being shaken by Lindland's arrogance and infuriated by what he saw as the mutually enjoyable alliance between Lindland and Linda Wyetzner, lawyer for the relator. I told him, "If you and Linda can't stop kicking each other under the table and laughing, it's going to be atrocious." I told him that. It was like a party for the two of them."
20. When the case against him was settled, Charlie hoped to bring a case for defamation and violation of academic freedom against Northwestern University. But James McGurk, in a 2019 telephone interview with author, remembers pointing out to him that Northwestern's lawyers had made that impossible in the agreement that facilitated his transfer to South Carolina:

    "He got a new job in South Carolina, and that settlement has extremely broad waiver provisions. So, if Charlie were going to sue Northwestern, the first thing he'd have to be able to battle through is a multi-, exhaustively prepared, carefully documented termination agreement that says, "You will never sue us for anything ever again from the beginning of time to now for any possible anything."

21. "Confidential Settlement Agreement and Release," *United States of America v. Charles L. Bennett*, United States District Court, Northern District of Illinois, Eastern Division.
22. James A. McGurk, Esq. memorandum to Nancy Gunderson, Esq; Karl Alvarez, Esq; Tiffani Redding, Esq. August 8, 2014.
23. Professor Elizabeth Calhoun, interview with author, Palm Spring, CA, 2021. In her FBI summary interview, the FBI reported that they had subpoenaed her bank records. Perhaps they were looking for a safe haven for money Charlie supposedly stole—although they knew that Feyi's bank had already revealed the full amount.
24. Leaving Northwestern, Dr. Rosen is currently Provost and Chief Scientific Officer at the City of Hope Cancer Center.
25. David Meyer, telephone interview with author, July 31, 2018.
26. Amy Haseltine, "Notice to Suspend and Proposed Notice to Disbar Dr. Charles Lee Bennett," Suspension and Debarment Official and Deputy Assistant Secretary of the Department of Health & Human Services," May 19, 2015. On file with author. This thirty-eight-page letter was signed on page 4, making the reading complicated and frustrating.
27. Amy Bennett, written correspondence with attorney Rob Henning, 2015. Telephone interview with Amy Bennett, 2019.

Chapter 9

1. The West German pharmaceutical company, Chemie Grunental Gmb H patented the medicine as an anticonvulsive drug, and then discovered its sleepy, relaxed consequences. Shuang Zhou et al., "A Notorious Sedative to a Wonder Anticancer Drug," ncb.nlm.nih.gov/pmc/articles/pmc4112512/.
2. In 1960 the courageous head of the FDA, Dr. Frances Kelsey, insisted to the Investigational Drug Branch of the FDA, that the drug should be kept out of the United States. In 1962, Pres. John F. Kennedy awarded her the highest civilian honor of the Presidential Award for Distinguished Federal Civilian Service. "Biography: Dr. Frances Kathleen Oldham Kelsey," cfmendicine.nlm.nih.gov/physicians/biography_182.html Visited February 2022.
3. Jerusalem's Hadassah University Hospital discovered that thalidomide reduced a leprosy sufferer's symptoms. http://broughttolife.sciencemuseum.org.uk/broughttolife/themes/controversies/thalidomide Visited March 2020.
4. Waqas Rehman et al., "The Rise, Fall and Subsequent Triumph of Thalidomide: Lessons Learned in Drug Development," *Ther Adv Hematol* 2(5) (Oct. 2011): 291–388; www: ncbi.hlm.nih.gov/pmc/articles/PMC3573415/ Visited February 2022.
5. *United States of America et al v. Celgene Corporation*, Docket No. 2:10-cv -03165 (C.C. Dal. App. Apr 27, 2010), Court Docket.

6. Putative authors are KOL (Key Opinion Leaders) and are held out as original authors. Sergio Sismondo, Ghost- Managed Medicine: Big Pharma's Invisible Hands (Mattering Press, 2018): 15. Articles written by pharmaceutical companies (outsourced to communication companies) are published under the names of doctors and professors who received payment. These doctors and professors are frequently paid for the article and for any speeches extolling the medicines. Sismondo: 17.
7. A citizen petition is a process afforded under Section 10.30 of Title 21, Volume 1, of the Code of Federal Regulations. It permits any person to request the FDA Commissioner to "issue, amend, revoke a regulation or order or take or refrain from taking any other form of administration action" over which the commissioner has statutory authority.
8. The 2020 website warnings about Revlimid are sobering: "Severe pregnancy harm. Revlimid isn't recommended for pregnant women. This is because Revlimid contains the drug lenalidomide, which is very similar to the drug thalidomide. In pregnant women, thalidomide can cause the death of a fetus or severe birth defects . . . Severe decrease in blood cell levels. Revlimid can cause neutropenia (low levels of white blood cells). It can also cause thrombocytopenia (low levels of platelets, another type of blood cell). These conditions can raise your risk of getting infections. These conditions may also make you bruise or bleed more easily . . . Dangerous blood clots.

   Venous thromboembolism is a blood clot in a vein. Arterial thromboembolism is a blood clot in an artery. Deep vein thrombosis (DVT) and pulmonary embolism are types of thromboembolism. Taking Revlimid can significantly raise your risk of developing DVT and pulmonary embolism. Also, if you take Revlimid along with dexamethasone for multiple myeloma, DVT or pulmonary embolism can lead to heart attack or stroke." https://www.medicalnewstoday.com/articles/326466 Visited March 2020.
9. See chapter 4 for discussion and some consequences of qui tam suits.
10. "Thalomid (thalidomide). "Thalomid® is an oral immunomodulatory drug, an agent that can modify or regulate the immune system. It has both anti-inflammatory and anti-cancer activities. Thalidomide was first used to treat multiple myeloma in 1997 and officially approved for use in treating myeloma in combination with the steroid dexamethasone in 2006. The first effective new drug to treat MM in decades, Thalomid launched a new era of "novel therapies." It gave rise to a next generation of immune modulators with increased efficacy and reduced side effects, or the drugs Revlimid® (lenalidomide) and Pomalyst® (pomalidomide).

    Currently in the United States, Thalomid is less commonly prescribed than its successor, Revlimid. However, Revlimid can cause low blood counts, and Thalomid has a smaller impact on the bone marrow's ability to make new blood cells. Thalomid may be a good alternative to Revlimid for patients with low blood cell counts." https://www.myeloma.org/treatment/current-fda-approved-medications/thalomid-thalidomide Visited March 2020.

11. Brian Melley, AP, "Celgene Settles Whistleblower Suit for $280M over fraudulent use of Thalomid," *USA Today*, July 26, 2017. One wonders why it took the altered code instruction after her many years working for the company to recognize the deception involved in her role. Her story of outraged innocence seems notably self- justifying. Possibly she learned about and was at least partly motivated by large sums of money achievable under a False Claims Act.
12. A full discussion of off-label abuses can be found in Marcia Angell's *The Truth About Drug Companies: How They Deceive Us and What to Do About It*, (Random House, 2004). The theory behind "off label" is that drug companies cannot market drugs for uses not already approved by the FDA. "Doctors, however, are not constrained by this law. They are permitted to prescribe drugs for whatever uses they want. So if a drug company can somehow convince doctors to prescribe drugs for off-label uses, sales go up." 137. That is how Celgene was skirting the law: using sales reps to encourage use outside FDA approvals. (It is illegal to offer doctors kickbacks (bribes) to prescribe drugs, but the pharmaceutical industry has developed an enormous web of rewards that skirt "legal.") Angell: 128.
13. Justin Brooks, email to author, August 2018.
14. Defendants' Motion for Summary Judgment 2016 on file with author: "The Complaint is devoid of allegations specifying what purportedly false claims for reimbursement were submitted to the government, how they were false, or even who submitted them. Moreover, whether a claim for off-label drug use is reimbursable turns not on how the drug was promoted, but on whether its off-label use was 'medically accepted.' Here, Brown fails to address or acknowledge that many of the off-label uses at issue were medically accepted and therefore required to be reimbursed by the government. Nor does Brown adequately or plausibly allege facts indicating that physicians prescribed these drugs (resulting in the submission of claims to the government) simply because Celgene improperly promoted them—and not based on their independent medical judgment and training, the available medical literature, and the needs of their patients."
15. Charles Bennett, Expert Statement in *Brown v. Celgene Corp*, 226 F.Supp 1032 (S.Cal 2016). On file with author.
16. In attorney Stephen Sheller's carefully documented research into the flow of dangerous drugs, *Big Pharma, Big Greed* (Big Arm Press, 2019), he reports that "Excerpta Medica was paid to manage the content, distribution, and timing of data dissemination [for a Johnson & Johnson drug] to competitive advantage. . . . Thus the contents and objectives of the articles were outlined and text written before they even found supposed authors." :72–71.

17. United States District Court, Central District of California, Western District. Case # 10-cv-03165 (GHK) SSx. "Joint Stipulation Regarding Defendant Celgene's Motion for Summary Judgment or, in the Alternative, Partial Summary Judgement." Oct. 16, 2016.
18. Plaintiff's Response to Defendant's Motion for Summary Judgement.
19. Citing e.g., Ebeid ex rel. *U.S. v. Lungwitz,* 616 F.3d 993, 1001 (9th Cir. 2010); *Mikes v. Straus,* 274 F.3d 687, 700 (2d Cir. 2001).
20. Brian Melby, "Celgene Settles Whistleblower Suit for $280 Million over Fraudulent Use of Thalomid." USAToday.com. 2017/07/16. www.usatoday.com/story/money/2017/07/26/celgene-settles-whistleblower-suit-280-m-over-fruadulent-use-of-thalomid/511792001/
21. Justin Brooks. Telephone interview with author, August 2018.

Chapter 10

1. Bennett Holman and Kevin C. Elliott, "The promise and perils of industry-funded science (March 2018), *Philosophy Compass*. 2018;e12544; https://doi.org/10.1111/phc3.12544.
2. As you have just read, "The discovery of ciprofloxacin was an important medical breakthrough and it opened the door for further research, development, and marketing of new class of antibiotics. Ciprofloxacin was the first fluoroquinolone brought to the market. It was discovered in 1981 by Bayer, the German-based drug and chemical company. In 1987 Cipro® was approved by the FDA in the United States as the first oral broad-spectrum antibiotic of this class." Emedexpert.com. Last visited March 2020.
3. *The Cancer Letter*, 33 #21, June 1, 2007:1.
4. Ben Djulebeovic. Telephone interview with author 2019. Djulebeovic recalls that Charlie discussed a threat with him before the presentation.
5. The e-mails demonstrate that Amgen followed Dr. Bennett's presentations very closely. AUSA Peter Winn has other e-mails from Amgen that demonstrate the intent of Amgen to discredit Dr. Bennett and to end his academic career.
6. John Glaspy, MD, email to Amgen executive Trish Hawkins, June 4, 2007. On file with author.
7. Amgen email 10/1/08. On file with authors.
8. Kelley Davenport, to 14 Amgen persons, February 26, 2008. On file with author. Pearlmutter then became the non-executive director at Merck Labs. He now heads a start-up, Eikon Therapeutics, pinpointing individual molecules within human cells. "Roger Perlmutter, former Merck R&D head, becomes CEO of startup". *STAT*. 2021-05-05. Retrieved 2021-05-06.
9. David Brier email to "Top Amgen Officials, March 6, 2008, recovered in Wynn's class-action investigation and sent to Bennett. On file with author.

10. McGurk, telephone call with author, June 5, 2017.
11. "People with neutropenia have an unusually low number of cells called neutrophils. Neutrophils are cells in your immune system that attack bacteria and other organisms when they invade your body. Neutrophils are a type of white *blood* cell. Your bone marrow creates these cells. They then travel in your bloodstream and move to areas of infection where they ingest and then neutralize the offending bacteria." https://www.webmd.com/a-to-z-guides/neutropenia-causes-symptoms-treatment#1 Visited March 2020.
12. Horwitz emails and multiple phone interviews with author 2019.
13. Gardiner-Caldwell Communications (GCC). GCC is an agency within Healthacare Communications Network (HCN). HCN is a division of KnowledgePoint 360 Group, LLC. They point out that they also rely on Scientia Scripta for publication communication.
14. "GOG-101 study," 18 Eur. *Gynecol. Oncol.* 25 (February 2008). 133–144. Written by GCC, signed by Dr. Dale. Crawford, Dale, Lyman, "Chemotherapy-Inducing Neutropenia: Reviewing Consequences and New Directions for Its Management," 25 *Cancer* (2004): 228–237.
15. Celgene had also used Exerpta Medica to create bogus articles purportedly written by scientific experts. In that case, "Beverly Brown and others were awarded approximately $300 million in a whistle-blower case." *USAToday*.com, July 16, 2017.
16. E.g., Fugh-Berman, A and Haris, "Following the Script: How Drug Reps Make Friends and Influence Doctors," 4 (4) PLosMed (April 2007):4; "POGO Letter to NIH on Ghostwriting Academics Project on Government Oversight," March 24, 2012. www.pogo-files/letters/public-health/ph-iis-20101129.html.
17. Department of Justice Office of Public Affairs, Wednesday, December 19, 2012.
18. The DOJ agreed to not mention ghostwriting. For example, this Dec. 19, 2012, release from the Dept. of Justice:

    Biotech Giant Pleads Guilty to Illegally Introducing Drug into Market for Uses That the FDA Declined to Approve; Will Pay $612 Million to Resolve False Claims Act Suits and $150 Million in Criminal Penalties and Forfeiture.

    The plea is part of a global settlement with the United States in which Amgen agreed to pay $762 million to resolve criminal and civil liability arising from its sale and promotion of certain drugs. The settlement represents the single largest criminal and civil False Claims Act settlement involving a biotechnology company in US history.

    https://www.justice.gov/opa/pr/amgen-inc-pleads-guilty-federal-charge-brooklyn-ny-pays-762- million-resolve-criminal Visited March 2020.

19. Dr. Dale's projects continue to focus on understanding the mechanisms and finding therapies for hereditary hematological diseases causing neutropenia. gim.uw.edu/research/david-dale-md. Visited February 2022.
20. Pharmaceutical companies bribed FDA investigators, including FDA "investigator" Charles Chang, who accepted numerous bribes to allow faulty generic drugs into the United States. Eban: 222–224. "Dingell's committee uncovered corruption that seemed to have no bottom. Generic drug executives had roamed the FDA's halls, dropping envelopes stuffed with thousand in cash onto the desks of [generic drug] reviewers. Charles Chang had accepted numerous bribes. . . . Rep. Ron Wyden (D-OR) called the generic drug industry "a swamp that must be drained." See Katherine Eban, 223.
21. News stories continue to reveal the opioid crisis. Purdue Pharmaceutical has pled guilty to selling off-label and to not keeping track of their distribution of ramped-up manufacturing, distribution, and sales of oxycodone and hydrocodone. Their market share was a mere 3.3 percent. SpecGx (Mallinckrodt) accounted for 37.7 percent for "an astonishing 76 billion prescriptions of opioids between 2006 to 2021." Geoff Mulvihill and Larry Fenn published a damning article in the *St. Louis Post-Dispatch* that revealed the craven acknowledgment of opioid sales. A drug distributor and Mallinckrodt salesman admitted that addictive opioids were sold "just like Doritos. Keep eating, we'll make more."

    stltoday.com/news/local/crime-and-courts/mallinckrodt-and-other-generic-drugmakers-sold-most-opioids-during-overdose-crisis/article_c8eea569-5757-5b67.
22. Bennett and his teams continue to alert his colleagues and the public to the avalanche of retaliations by drug companies. McCoy et al., Caveat Medicus: Clinician Experience in Publishing Reports of Serious Oncology-Associated Adverse Drug Reactions." 23 #24, J. Clin. Oncol (Dec. 2005): 8894–8905. dx.plos.org/10.1371/journal.pone.021521 or PLOS ONE 14 (7).
23. Dr. John Thompson, Dr. Patricia A. Baird, Dr. Jocelyn, *Ecclectica*, "The Olivieri Case: Context and Significance," (December 2005). ecclectica.brandonu.ca/issues/2005/3/Article-2.html.
24. Olivieri's funding 1989–1996 came from the Medical Research Council of Canada and then from 1993 forward from Apotex, until it wasn't. It was not licensed in the United States until 2011; Canada, 2015.
25. Francoise Baylis: "Defending academic freedom did not require taking a stance on whether Dr. Olivieri was right or wrong in her assessment of the efficacy of Deferiprone. It merely required defending her right to present her views to her scientific peers." www.jme.bmj.com/content/30/1/44.
26. Dr. Nancy Olivieri, "Nothing is Right about the Approval of Aducanumab and Nothing's New," *The BMJ Opinion,* Nov. 4, 2021, blogs.bmj.com/bmj/2021/11/04/nothing-is-right-about-the-approval-of-aducanumab-

and- nothings-new/ "The FDA commissioned its Department of Scientific Investigation to conduct an inspection of my original trials. Over a week, the FDA inspector conducted a painstaking inspection of all my clinical data, confirmed concerns about the safety of deferiprone that I'd been struggling to raise for years and showed that Sherman's allegations were unfounded. After examining all the data, the inspector reported that with respect to my trials' primary (non-surrogate) endpoint of effectiveness, liver iron concentration, Sherman had submitted data to the FDA that had excluded 45% of subjects.

The most immediate outcome of this conclusion was that the FDA rejected Sherman's application. The FDA also informed Sherman that, given that my trials had been terminated prematurely, there existed no adequate studies of effectiveness for deferiprone. The FDA therefore demanded at least one additional prospective, randomized, controlled clinical study to verify the effectiveness of deferiprone.

Sherman demanded that the FDA approve the drug based on selected data from older studies which, as in the case of aducanumab, had used a surrogate marker of "effectiveness." Ignoring its own written findings, the FDA issued "accelerated approval" for deferiprone (albeit as "last resort" therapy.) Asked whether the FDA had ever issued approval based on data of such poor quality, an FDA official responded as follows: "Not that I am aware of. I want to make sure this doesn't establish a precedent."

27. Baylis, F. The Olivieri Symposium. The Olivieri Debacle: where were the heroes of bioethics? https://jme.bmj.com/content/30/1/44.
28. Leonid Schneider, *For Better Science* reported that the University of Toronto and Apotex had negotiated for a $92 million donation, which was withdrawn. forbetterscience.com/2020/06/29/motherisk-crook-gideon-koren-now-at-ariel-university. Also see: Francoise Baylis, "The Olivieri Debacle: Where Were the Heroes of Bioethics?," jmedethics.com, June 17, 2003; "Heroes in Bioethics," 30 (3) *Hastings Center Report* (2000): 34–39; "Report clears researcher who broke drug company agreement," https://www.ncbi.nlm.nih.gov/pmc/articles/PMC1121590. Also see: Wendy Wagner and Tom McGarity, "The Art of Concealing Unwelcome Information," *Bending Science,* 97–101, (Harvard University Press, 2008).
29. As we investigated this strange, underreported segment of the medical world, we learned that Apotex's Chief Executive Officer Barry Sherman and wife Sunny were murdered in their Toronto home in 2017. "The Unsolved Murder of an Unusual Billionaire," Bloomberg.com/features/2018-apotex-billionaire-murder.
30. Francoise Baylis, "Stories of Silence," www.jme.bmj.com/content/30/1/44.
31. "Boots contacted UCSF administrators, including Chancellor Joseph Martin, and several department heads, charging that Dong's study had been flawed in numerous ways —patient selection and compliance, assay reliability, method

of determining bioequivalence, statistical analysis. The company even raised the spectre of conflict-of-interest and ethical violations on Dong's part. UCSF responded by conducting two investigations of Dong's work, one by investigational drug pharmacist Scott Fields, who concluded in June 1992 that any flaws in her study were "minor and easily correctable," and one by Benet, who found that the study had been rigorous and that Boots's criticisms were "deceptive and self-serving." Karen Kerr, "Drug Company Relents, Dong's Findings in *JAMA*," *The Synapse,* May 22, 1997. synapse.library.ucsf.edu/?a=d&d=west19970522-01.

32. Dr. Dong. Email response to author, September 11, 2021.
33. Stephen Fried, *Bitter Pills*, 341–343.
34. February 1996. Also see Lawrence Altman provided a prominent expose. *The New York Times (*April 16, 1997): Section A, p.1.
35. *JAMA* 4 (no.31), April 1997.
36. *JAMA* 4 (no. 31), April 16, 1997: 1. Editorial for this issue included: "The issue [Dr. Dong's] also contains a lengthy editorial by Drummond Rennie recounting Dong's ordeal. Rennie concludes by citing another *JAMA* article by D. Blumenthal et al. reporting that almost 20% of 2100 investigators in the life sciences had experienced delays of over six months in the publication of their results. Some 28% of respondents thought the purpose of the delays was "to slow dissemination of undesired results." "I'm not the only one to whom this has happened," Dong told Synapse. "I would warn anyone who plans to conduct research that is funded by private companies to not accept contracts with those limitations Unless we stick to our convictions to share the honest results of our work, this enforced silence will continue, and that is wrong."
37. Ben Goldacre, *Bad Pharma: How Drug Companies Mislead Doctors and Harm Patients*, Farber and Farber 2012): 91–93.
38. Report and Dr. Buse's testimony Feb. 20, 2010. Sen. Charles E. Grassley. "Ghost writing in the medical literature," 111th Congress, United States Senate Committee on Finance," on file with authors; also see finance.senate-gov/release/Grassley-baucus-release-committee-report-on-avandia.
39. Also see Golomb, Dr. Beatrice., "Beyond Belief: Your brain on medical politics," YouTube video, 19.05 min., 2008.golombresearchgroup.org/videos.
40. Fabio Turone, "MSD Italy is criticized for threatening legal action over prescription advice to GPs', July 4, 2014, BMJ 2014;349:4441.
41. Colin Meek, "Spanish journal wins lawsuit over COX-2 drug editorial (April 2004), researchgate.net/publication/8677078. Original discussion of suit, Lisa Gibson, "Drug company sues Spanish Bulletin over fraud claim," BMJ. 2004 (Jan. 24: 328(7433): 188. Perhaps Merck capitulated because that same year Money Magazine revealed its official, hidden emails stating that Merck had worried about the risks of Vioxx beginning back in the 1990s.

money.cnn/com/2004/11/01/news/fortune500/Merck. Editorial: "Vioxx: an unequal partnership between safety and efficiency" (364) October 9, 2004, The Lancet.com, 1288.

42. General practitioner Dr. Andrew Gunn refused to apologize for "insulting" The University of Queenland's "corporate sponsor." Gardasil, developed in collaboration with the university/pharmaceutical company, was directed at girls and women 9–26 to help avoid HPV (human papillomavirus). The university asked Dr. Gunn for a public apology to the Australian distributor. It did not go well. When academics and scientists pointed to the out-of-line criticisms, the university admitted error and withdrew its complaint. Peter McCutcheon, "University admits error over forced Gardasil apology," May 15, 2008.
43. Today's news headlines add fuel to that fire with their coverage of the pharmaceutical bribery schemes to get doctors to overprescribe opioids. For example, John Kapoor and four others were convicted of racketeering conspiracy; the Sackler family faces lawsuits for OxyContin bribes to doctors; Insys Therapeutics used sales executives to hire exotic dancers for physicians at conferences promoting Subsys, a fentanyl-based oral spray. See, for instance, "Ex-drug company execs face reckoning in bribery case," *Austin American-Statesman,* January 13, 2020: B5; "Congressman Chris Collins to prison for leaking confidential pharmaceutical company's failed drug trial," syracuse.com/state/2020/08.
44. In 2004, Dr. Gurkirpal Singh of Stanford University testified before the United States Committee on Finance. The record shows that an executive at Merck had intimidated/warned him if "he persisted in his request for data on Merck's drug, Vioxx." Repeated and discussed in the Dr. Buse's testimony 2007 [see footnote #38 above].
45. Eban, *Bottle of Lies,* xiv.
46. Congress is also implicated in the greed surrounding the profitable drug companies. US Rep. Chris Collins was sentenced to two years in federal prison for leaking confidential information about a failed drug test so that he could avoid $800,000 in stock market losses. www.syracuse.com/state/2020/08.
47. For a full discussion of my involvement in faculty committees' investigations into academic freedom, see Julius Getman, *In the Company of Scholars* (University of Texas Press, 1992.).
48. https://www.justice.gov/usao-ndil/pr/northwestern-university-pay-nearly -3-million-united-states-settle- cancer-research- July 30, 2013grant#:~:text=Charles%20L.%20Bennett%2C%20et%20al.%2C%20No.%2009%20C,States%20Attorney%20for%20the%20Northe rn%20District%20of%20 Illinois.

49. None of the injustice is mitigated by the fact that Charlie was able to achieve professional success and public acclaim as a whistleblower to the dangers of fluoroquinolones.
50. "Triple (treble) damages are a tripling of an award in a lawsuit against a defendant who is subject to punitive damages. Punitive damages are awarded when a defendant's conduct was malicious, or in reckless disregard of plaintiff's rights. Punitive damages are often awarded in cases of fraudulent acts by the defendant." https://definitions.uslegal.com/t/triple-damages/ Visited March 2020.
51. In *Universal Health Services, Inc. v. United States ex rel.* Escobar, 579 U.S. 14 (2016).
52. The [Escobar] Court said what makes the False Claims Act applicable is "when a defendant submitting a claim makes specific representations about the goods or services provided, but fails to disclose noncompliance with material statutory, regulatory, or contractual requirements that make those representations misleading with respect to those goods or services." *Escobar*: 2. Also see: https://www.whistleblowers.org/amicus-curiae-briefs/universal-health-services-v-u-s-ex-rel-escobar-supreme-court-victory-for-whistleblowers/Visited March 2020.

Chapter 11

1. Allison Gandey, "European Regulators Say Transfusions Safer Than ESAs in Cancer," June 30, 2008, https://medscape.com/viewarticle/576864.
2. Allsion Gandey, "Erythropoiesis-Stimulating Agents May Not Be Safe for Cancer Patients (Aug. 2, 2000) Medscape.org/viewarticle/570805.
3. A 2022 *Tampa News* report ties some antibiotics to recent and puzzling suicides (see Appendix E).
4. Crystal Britt, "Potentially dangerous antibiotics," (Mother uneasy about taking new drug levofloxacin for respiratory tract infection), Feb. 20, 2019. Kfvs.12.com/2019/02/20/potentially-dangerous-antibiotics/.
5. Steve Rosen, interview with authors, fall 2021. https://www.southcarolinapublicradio.org/show/south-carolina-business-review/2022-08-01/smartstate-chair-identifies-medication-safety-protocol-through-twitter.
6. Charles L. Bennett, William Kennedy Smith, and Eric D. Perakslis, "How the government failed us on opioids," *Los Angeles Times,* August 8, 2019: A11. "A good starting point for reforming the system would be increases transparency about drugs already recognized as particularly dangerous. These drugs, currently numbering about 70 (including opioids), carry the FDA's so-called 'black box warning' intended to alert patients and their doctors to the high risks associated with the drugs. But that is not enough. We propose a 'black-box' data base or 'registry,' publicly available and simple to use . . . ."

7. C. L. Bennett, K. Gundabolu, L. W. Kwak, B. Djulbegovic, O. Champigneulle, B. Josephson, L. Martin, and S. T. Rosen, "Using Twitter for the Identification of COVID-19 Vaccine-Associated Haematological Adverse Events," *Lancet: Digital Haematology*, 9 (1), Jan. 01, 2022: e12-e13.
8. Shamia Hoque, Brian J. Ci et al. *PLOS ONE* (June 25, 2020) PLOSONE: doi.org/10.1371/journal.pone.023454.. Also see "End of an Era: Erythropoiesis-Stimulating Agents in Oncology," *International Journal of Oncology* 146 (2020): 2829–2835.
9. Charles L. Bennett, N. Olivieri, S. Hogue, et al., "David versus Goliath: Pharma and academic threats to individual scientists and clinicians," June 24, 2022, JoSPI.

Appendix A

Feyifunmi Sangoleye, June 25, 2013, criminally charged with Violation of Title 18: U.S. Code, Section 641:

FEYIFUNMI SANGOLEYE,

1. Defendant herein, knowingly did embezzle, steal and convert to her own use, and to the use of another, money of the United States, namely a check in the amount of $6,000, dated July 29, 2008, drawn on a bank account of Northwestern University and made payable to ATSDATA, which funds were disbursed as part of the National Institutes of Health's RADAR multiple myeloma grant to Northwestern University, which money defendant was not entitled to receive;
2. Between on or about May 15, 2007, and on or about June 17, 2008, SANGOLEYE created approximately ten false invoices that purported to have been issued by an entity called ATSDATA for consulting services it allegedly performed in connection with various

Northwestern University grants, including the RADAR grants. ATSDATA did not exist and never performed consulting services in connection with any grant.

3. SANGOLEYE created and caused to be created, on Northwestern University's computerized accounting system, a "vendor code" that was assigned to ATSDATA. SANGOLEYE submitted and caused to be submitted the approximately ten false ATSDATA invoices she had created, using the ATSDATA vendor code, to the university for processing and payment.
4. SANGOLEYE rented and caused to be rented a Post Office box in Chicago to receive checks written to ATSDATA by Northwestern University. She established and caused to be established a checking account at Citibank in the name of ATSDATA for use in negotiating the checks she fraudulently received from the university, and for the purpose of converting the proceeds of those checks to her own use and to the use of another.
5. Beginning no later than June 29, 2007, through on or about July 29, 2008, Northwestern University issued eight checks totaling $86,000 to ATSDATA. The checks represented disbursements for various medical research grants, including the RADAR grants. Knowing that ATSDATA did not exist, and that it had never performed any consulting services in connection with any grant, SANGOLEYE obtained the eight checks, and caused them to be deposited to the ATSDATA account at Citibank. SANGOLEYE converted the proceeds of the ATSDATA checks to her own use and to the use of another.
6. On or about July 31, 2008, at Chicago, in the Northern District of Illinois, Eastern Division, and elsewhere, In violation of Title 18, United States Code, Section 641.

Appendix B

02-06-03 NORTHWESTERN UNIVERSITY WILL PAY $5.5 MILLION TO RESOLVE FALSE CLAIMS ACT AND COMMON LAW ALLEGATIONS (justice.gov). oig.nasa.gov/docs/pr2009-B.pdf

FOR IMMEDIATE RELEASE	**CIV**
THURSDAY, FEBRUARY 6, 2003	**(202) 514–2007**
WWW.USDOJ.GOV	**TDD (202) 514–1888**

NORTHWESTERN UNIVERSITY WILL PAY $5.5 MILLION TO RESOLVE FALSE CLAIMS ACT AND COMMON LAW ALLEGATIONS

WASHINGTON, D.C.—Northwestern University will pay the United States $5.5 million to settle allegations that the school violated the False Claims Act with regard to claims in connection with federally-sponsored medical research grants, the Justice Department announced today. The government alleged that Northwestern misled the United States into paying more money than the Chicago-area school was lawfully entitled to receive.

For example, the government alleged that in completing applications for the National Institutes of Health and other federal agencies grants, Northwestern overstated the percentage of its researchers' work effort that they were able to devote to the grant. The United States also alleged that the university knowingly failed to comply with federal government requirements that a specified percentage of the researchers' effort be devoted to the grant.

"This settlement illustrates the importance to the United States of ensuring that universities and other institutions make proper use of federal research funds," said Assistant Attorney General Robert D. McCallum, Jr., head of the Department's Civil Division.

The civil settlement includes the resolution of claims brought under the qui tam or whistleblower provisions of the False Claims Act against Northwestern by Richard Schwiderski, a former employee of the University's Office of Research Sponsored Programs. The suit was filed in the federal court in Dallas, Texas in March 2000, but was later transferred to U.S. District Court for the Northern District of Illinois in Chicago.

As a result of today's settlement, Mr. Schwiderski will receive $907,500 of the total recovery as his statutory award. Under the qui tam provisions of the False Claims Act, a private party can file an action on behalf of the United States and receive a portion of the settlement if the government takes over the case and reaches a monetary agreement with the defendant.

The Justice Department received analytical support in this case from the National Institute of Health's Division of Program Integrity in Rockville, Maryland and the United States Department of Health and Human Services' Office of the Inspector General (HHS-OIG) in St. Paul, Minnesota. The case was investigated by HHS-OIG in Chicago,

Illinois. The case was handled by the Civil Division of the Department of Justice with assistance from the US Attorney's Office in Chicago.

03–076

Appendix C

NIH Grants Policy Statement, Part 1,1.1 Abbreviations defines as follows:

Grantee: (Northwestern University and the Robert Lurie Cancer Center) The organization or individual awarded a grant by the NIH that is responsible and accountable for the use of the funds provided and for the performance of the grant supported project. . . . The Grantee is the entire legal entity even if a particular component is designated NoA. The Grantee is legally responsible and accountable to the NIH for the performance and financial aspects of the grant supported project or activity.

Principal Investigator: (Dr. Bennett) The individual designated by the applicant (Northwestern University) to have the appropriate level of authority and responsibility to direct the project to be supported by the grant. . . . PIs have the authority and responsibility for leading and directing the project, intellectually and logistically.

SF 424 Application Guide: relates to the duties of the Grantee (Northwestern University and the Robert Lurie Cancer Center) relating to the grant application process. It also defines Grantee: **Grantee:** (Northwestern University and the Robert Lurie Cancer Center) The organization or individual awarded a grant . . . by the NIH that is responsible and accountable for the use of the funds provided and for the performance of the grant supported project. . . . The Grantee is the entire legal entity even if a particular component is designated in the NoA. The grantee is legally responsible and accountable to the NIH for the performance and financial aspects of the grant supported project.

Appendix D

McGurk tried to keep Amy abreast of the multiple parties and filing:

From: James A. McGurk
To: Amy
Sent: Mon, Jun 16, 2014 3:24 pm
Subject: Re: Bennett

Dear Amy:
First, the current status:
The government counsel, AUSA Kurt Lindland, advised me that the "client", that is the Department of HHS and NIH, was prepared to agree with a settlement here in U.S. District Court without any agreed upon "debarment" for Charlie. AUSA Kurt Lindland said that the HHS and NIH lawyers said that they would proceed with an administrative proceeding to debar Charlie for a much longer time period than three years.

I told Kurt Lindland that we would attempt to deal with HHS and NIH directly so that a settlement without the debarment was fine. I then said that Charlie could not pay $500,000 and he certainly could not pay $13,000 plus per month.

Lindland said we were going back on an agreed upon deal. I denied that we had a deal.

Lindland has filed a motion to enforce a settlement which he contends is the $500,000, based upon the government's dropping the debarment.

Appendix E

Jackie Callaway, "Growing Number of Suicides Linked to Popular Antibiotics," (Feb. 3, 2022) https://www.abcactionnews.com/news/local-news/i-team-investigates/popular-antibiotics-linked-to-growing-number-of-suicide-deaths-patients-unaware-of-side-effects. (quoting Charlie on earlier attempts to persuade the FDA to include suicide on black box warnings).

Popular antibiotics linked to growing number of suicide deaths, patients unaware of side effects:
Experts say drug may be tied to thousands of deaths

"Reports of deadly side effects are on the rise in one of the most widely prescribed antibiotics on the market . . . the popular and powerful antibiotic Cipro diagnosed for pneumonia."

. . . "Reports filed with the FDA tie 219 suicides to Cipro and Levaquin, two popular antibiotics used to treat serious infections."

. . . Cipro's warning about suicide as a side effect is found, buried, in **the Cipro drug insert and found anxiety, insomnia and suicide warnings on page 12.**

Charles Bennett, Medical expert, continues to petition FDA for checks on the popular antibiotic to protect public from suicide risk. He suspects that only 1–10% of all adverse drug side effects are ever reported.

A nationally recognized medication safety expert, Dr. Charles Bennett, petitioned the FDA in 2014 and again in 2019 for increased suicide warnings for Cipro and Levaquin. They're in a class of drugs known as fluoroquinolones.

In 2018 the FDA issued its strongest "black box" warnings for these drugs. It states that the antibiotics are "associated with disabling and potentially permanent side effects of the tendons, muscles, joints, nerves and central nervous system."

The agency denied Dr. Bennett's petition to add suicide as part of the boxed warning, leaving it in a subsection on page 12 of the 52-page drug insert included with the prescription.

Dr. Bennett said physicians often don't know about the side effects associated with these drugs when they prescribe them. In an attempt to raise worldwide awareness, Dr. Bennett told the I-Team that he started work on a petition asking the CDC to recognize Fluoroquinolone Long Term Toxicity as a condition that can be billed to insurance.

Meanwhile, the makers of Levaquin stopped making the drug, citing other options for treatment. This week Cipro's maker, Bayer, responded to ABC Action News that " . . . *risks are communicated to physicians and patients in FDA-approved product labeling. . . . "The health and safety of patients who use Bayer products is our top priority.*

Here is additional TV coverage. Jennifer Kovaleski (Apr 04, 2022). Doctors, patients say more needs to be done to warn patients of side effects of popular antibiotics (thedenverchannel.com). "DENVER—Some of the most popular drugs prescribed to treat bacterial infections can cause serious side effects and are overprescribed, according to studies and reports.

The Food and Drug Administration has said that Cipro (ciprofloxacin), Ofloxacin and Levaquin (levofloxacin)—prescription drugs known as fluoroquinolones—can have rare but serious side effects. Those side effects include tendon rupture, neuropathy, impacts to the central nervous system and suicide.

"'It destroyed my life,' said Jon Horne, who was prescribed Ofloxacin."

Index